This is a copy of the publication:

Schüßler A, Walker C (2010)
The *Glomeromycota.* A species list with new families and new genera.
Arthur Schüßler & Christopher Walker, Gloucester. Published in December 2010 in libraries at The Royal Botanic Garden Edinburgh, The Royal Botanic Garden Kew, Botanische Staatssammlung Munich, and Oregon State University.

An electronic version of the copy is freely available online at www.amf-phylogeny.com

This version is identical to the original, printed publication cited above, including the errors. Therefore, this version contains one additional, initial page as a corrigendum, giving corrections of the errors in the original publication.

Corrections, 2 FEB, 14 FEB, 19 JUL 2011. The corrections are highlighted in red.

p 7. **FOR** *Claroidoglomeraceae* **READ** *Claroideoglomeraceae*

p 10. **DELETE**
Glomus pulvinatum (Henn.) Trappe & Gerd. [as *'pulvinatus'*], in Gerdemann & Trappe, Mycol. Mem. 5: 59 (1974)
≡*Endogone pulvinata* Henn., Hedwigia 36: 212 (1897)

p 11. **AFTER** Botanical Code for formal descriptions after 1 Jan 1935 **INSERT**)

p 14. **BELOW** ≡ *Endogone macrocarpa* var. *geospora* T.H. Nicolson & Gerd., Mycologia 60(2): 318 (1968) **INSERT**
≡ *Glomus macrocarpum* var. *geosporum* (T.H. Nicolson & Gerd.) Gerd. & Trappe [as *macrocarpus* var. *geosporus*], Mycol. Mem. 5: 55 (1974)

p16. **ABOVE** *Sclerocystis coccogenum* (Pat.) Höhn., Sber. Akad. Wiss. Wien, Math.-Naturw. Kl., Abt. 1 119: 399 [7 repr.] (1910) **INSERT**
Sclerocystis clavispora Trappe, Mycotaxon 6(2): 358 (1977)
≡ *Glomus clavisporum* (Trappe) R.T. Almeida & N.C. Schenck, Mycologia 82(6): 710 (1990)

p 19. **FOR** *Rhizophagus irregulare* **READ** *Rhizophagus irregularis*

p 19. **FOR** *Rhizophagus proliferus* (Błaszk., Kovács & Balázs) **READ** *Rhizophagus proliferus* (Dalpé & Declerck)

p 28. **FOR** *Scutellospora arenicola* Koske Koske & Halvorson **READ** *Scutellospora arenicola* Koske & Halvorson

p 29. **FOR** *Scutellospora pernambucana* Oehl, Oehl, D.K. Silva, **READ** *Scutellospora pernambucana* Oehl, D.K. Silva,

p 30. **FOR** Genus name: *Racocetra* Oehl, F.A. Souza & Sieverd., Mycotaxon: 334 (2009) **READ** Genus name: *Racocetra* Oehl, F.A. Souza & Sieverd., Mycotaxon 106: 334 (2009)

p 35. **FOR** *Acaulospora mellea* Spain & N.C. Schenck, in Schenck, Spain, Sieverding & Howeler, Mycologia 76(4): 689
READ *Acaulospora mellea* Spain & N.C. Schenck, in Schenck, Spain, Sieverding & Howeler, Mycologia 76(4): 690

p 39. **FOR** *Entrophospora nevadensis* J. Palenzuela, N. Ferrol & Oehl, Mycologia 102(3): 627 (2010) **READ**
Entrophospora nevadensis Palenz., N. Ferrol, Azcón-Aguilar & Oehl, in Palenzuela, Barea, Ferrol, Azcón-Aguilar & Oehl, Mycologia 102(3): 627 (2010)

p 41. **FOR**
Generic type: *Pacispora chimonobambusae* (C.G. Wu & Y.S. Liu) Sieverd. & Oehl ex C. Walker, Vestberg & A. Schüßler, in Walker, Vestberg & Schüßler, Mycol. Res. 111(3): 255 (2007)
≡*Gerdemannia chimonobambusae* (C.G. Wu & Y.S. Liu) C. Walker, Błaszk., A. Schüßler & Schwarzott, in Walker, Błaszkowski, Schwarzott & Schüßler, Mycol. Res. 108(6): 717 (2004)
≡*Glomus chimonobambusae* C.G. Wu & Y.S. Liu, in Wu, Liu, Hwuang, Wang & Chao, Mycotaxon 53: 284 (1995)
READ
Generic type: *Pacispora scintillans* (S.L. Rose & Trappe) Sieverd. & Oehl ex C. Walker, Vestberg & A. Schüßler, in Walker, Vestberg & Schüßler, Mycol. Res. 111(3): 255 (2007)
≡*Glomus scintillans* S.L. Rose & Trappe, Mycotaxon 10(2): 417 (1980)
≡*Gerdemannia scintillans* (S.L. Rose & Trappe) C. Walker, Błaszk., A. Schüßler & Schwarzott, in Walker, Błaszkowski, Schwarzott & Schüßler, Mycol. Res. 108(6): 716 (2004)
=*Glomus dominikii* Błaszk., Karstenia 27(2): 37 (1988) [1987]
=*Pacispora dominikii* (Błaszk.) Sieverd. & Oehl, in Oehl & Sieverding, J. Appl. Bot., Angew. Bot. 78: 76 (2004)

Pacispora chimonobambusae (C.G. Wu & Y.S. Liu) Sieverd. & Oehl ex C. Walker, Vestberg & A. Schüßler, in Walker, Vestberg & Schüßler, Mycol. Res. 111(3): 255 (2007)
≡*Gerdemannia chimonobambusae* (C.G. Wu & Y.S. Liu) C. Walker, Błaszk., A. Schüßler & Schwarzott, in Walker, Błaszkowski, Schwarzott & Schüßler, Mycol. Res. 108(6): 717 (2004)
≡*Glomus chimonobambusae* C.G. Wu & Y.S. Liu, in Wu, Liu, Hwuang, Wang & Chao, Mycotaxon 53: 284 (1995)

p 41. **BELOW** *Pacispora robigina* Sieverd. & Oehl, in Oehl & Sieverding, J. Appl. Bot. (Angew. Bot.) 78: 75 (2004)
DELETE *Pacispora scintillans* (S.L. Rose & Trappe) Sieverd. & Oehl ex C. Walker, Vestberg & A. Schüßler, in Walker, Vestberg & Schüßler, Mycol. Res. 111(3): 255 (2007)
≡*Gerdemannia scintillans* (S.L. Rose & Trappe) C. Walker, Błaszk., A. Schüßler & Schwarzott, in Walker, Błaszkowski, Schwarzott & Schüßler, Mycol. Res. 108(6): 716 (2004)
≡*Glomus scintillans* S.L. Rose & Trappe, Mycotaxon 10(2): 417 (1980)
=*Pacispora dominikii* (Błaszk.) Sieverd. & Oehl, in Oehl & Sieverding, J. Appl. Bot., Angew. Bot. 78: 76 (2004)

p 43. **FOR** ≡*Glomus aurantium* Błaszk., Blanke, Renker & Buscot, Mycotaxon 90: 540 (2004) **READ** ≡*Glomus aurantium* Błaszk., Blanke, Renker & Buscot, Mycotaxon 90: 450 (2004)

p 43. **FOR** Genus name: *Otospora* Palenz., Ferrol & Oehl **READ** Genus name: *Otospora* Oehl, Palenz. & N. Ferrol

p 43. **FOR** Generic type: *Otospora bareae* Palenz., Ferrol & Oehl [as *'bareai'*] **READ** Generic type: *Otospora bareae* Palenz., N. Ferrol & Oehl [as *'bareai'*]

p 50. **FOR** *Ambispora granatensis* J. Palenzuela, N. Ferrol **READ** *Ambispora granatensis* Palenz., N. Ferrol

p 53. **FOR** (Morton & Redecker 2001; Kaonongbua 2010). **READ** (Morton & Redecker 2001; Kaonongbua et al. 2010).

Comment on the gender of the epithets in *Redeckera*.
In publishing the new genus *Redeckera*, in honour of Dirk Redecker, we treated the gender as neuter, thus giving the epithets as *pulvinatum*, *megalocarpum*, and *fulvum*. We had inadvertently missed the recommendation 20A.1(i) in the Botanical Code requesting that all such epithets should be made feminine, and we apologise for this. However, because the names have been formally published, the requirements of Article 62 apply, and the neuter gender must be retained.

CONTENTS

INTRODUCTION

For many years, the molecular phylogeny of the *Glomeromycota* (Schüßler et al. 2001) has been published in parts, and frequently updated on the webpage www.amf-phylogeny.com. We also provide the Species 2000 & ITIS Catalogue of Life (Schüßler 2010) with those data. However, the International Code of Botanical Nomenclature (ICBN) does not allow solely electronic publication of taxonomic novelties, so formal changes could not be implemented on this widely used information source. Moreover, because it was not possible to establish the true phylogenetic placement of *Glomus macrocarpum*, which is the type species of *Glomus*, a deeply revised taxonomy for the *Glomeromycota* was impossible. Without knowing the phylogenetic position of this species, we lacked the evidence to classify our long proposed groupings at familial (Schwarzott et al. 2001) and generic levels (Schüßler et al. 2011). We have recently established *Glomus macrocarpum* in pot culture and sequenced the SSU rRNA gene to allow us to anchor the position of this fungus and thus establish its natural phylogenetic position in relation to others in this 'genus'. We can now restructure the systematics of the order *Glomerales* (*Glomeromycota*) and also the three other orders in the *Glomeromycota*. In this work, we have listed all glomeromycotan species presently described. We have created new families and genera based on recent phylogenetic analyses, established in large parts by our own research, and we have categorised those species for which the molecular identity is still unknown as 'species of uncertain position' in the taxonomic hierarchy, and listed them under their original genus. Also, a number of epitypes are established, all of which are based on living cultures available for scientific research.

The *Glomeromycota* consists of fungi that are generally considered to be obligately symbiotic. Although probably correct, this is an assumption based on analogy with the species for which the biology is known. Such species have been shown either (in one instance) to have a *Nostoc* (*Cyanobacteria*) species as a symbiont (Schüßler 2002), or (in all other known instances) to form an intimate symbiosis, generally known as an arbuscular mycorrhiza (AM), with embryophytes (land plants). As well as vascular land plants, hornworts (Schüßler 2000) and liverworts (Fonseca & Berbera 2008) also may form AM. Many glomeromycotan species are known to form AM, but many others have been described from field collected specimens for which the nutritional state of the fungus is unknown. Some species have been established in pot culture in the past, but are no longer available as living material, and consequently have not been subjected to molecular analysis. A few have had genetic markers DNA-sequenced before the demise of the cultures, or from adequately determined field material, and thus can be placed phylogenetically, but many remain to have their true phylogeny uncovered.

Historically, most species in this phylum have been described and named from the morphology of their spores. These are produced in the main ectocarpically in the substratum, or in the roots of their host plants. Some produce spores in unstructured dense masses or in structured sporocarps at or near the surface of the soil, and it was these that were first observed and named. However, Morton et al. (1998) argued cogently that the spore is uncoupled from other parts of the organism, and if this is so, variation among spores will not necessarily reflect the true phylogeny. This has been proven since molecular methods have been available, and it has been shown that 'cryptic' speciation exists. In fact, it is evident that sometimes spore morphology may be almost indistinguishable among species in different families or even orders (Morton and Redecker 2001; Walker et al. 2007; Walker 2008; Gamper et al. 2009).

For the most part, the phylogenetic base used here is the analysis of the small subunit (SSU) rRNA gene, but we consider also data from the large subunit (LSU) rRNA gene and the internal transcribed spacer (ITS) region (comprising the 5.8S rRNA gene and the ITS 1 and ITS2). For simplicity we refer to these as the SSU, LSU, and ITS region. Moreover, we take account of β-tubulin sequence data published in Msiska and Morton (2008) and Morton and Msiska (2010). The SSU subunit is too conserved for resolving species level in the *Glomeromycota*, but is an important marker for robust phylogenies down to (sub-) genus resolution. Sequences, considered sufficiently good to determine species level are in the region of 1500bp in length, covering part of the SSU and LSU and all of the ITS-region (Krüger et al. 2009; Stockinger et al. 2009) but are not yet available for most AM fungi. Generally speaking, family and generic resolution can be determined by using the SSU or parts of the LSU, and species level resolution can be obtained by additionally using further LSU and the ITS-region data (Stockinger et al. 2010).

Many AM fungal species are placed in genera without conclusive evidence. Because this work is based on a natural classification from molecular phylogeny combined with convincing morphological evidence when available, we have excluded species for which we lack such evidence, but have retained them in their original genera, as 'species of uncertain position'. As a consequence, the phylogeny we offer is imperfect, but gradually it may be possible to move such species to their correct taxa, when living material that can confidently be assigned a name is discovered and DNA sequences of sufficient quality can be analysed with consequential study of phylogeny. We have tended to be conservative in our approach, and thus organisms that may appear from morphology to fall into well defined groups have sometimes been placed among the species of uncertain position, pending the production of further evidence. We provide a phylogenetic tree showing the clades with their associated genus names (page 5), which was computed by using a maximum likelihood method based on near full length SSU rRNA gene sequences. The numbers at the branches show the bootstrap support for the respective topologies.

Where possible, cultures of AM fungi cited are given an unique numerical identifier from a purpose designed database used by C. Walker. Such numbers may be notional if the culture is deduced from externally provided information in the literature or actual if the culture is known for certain to have existed. Notional numbers may include a series of subcultures if full details are unknown. The numbers have an Attempt number and an associated subculture number. An example is Attempt 1495-0, which is a first attempt to establish a culture with spores from a field-collected sporocarp as inoculum. Voucher specimens from the C. Walker collection in E are given a number preceded by W, e.g., W5288 was taken from Attempt 1495-0 on 04 Mar 2009. Herbarium abbreviations are from Thiers, B. Index Herbariorum: A global directory of public herbaria and associated staff. New York Botanical Garden's Virtual Herbarium. http://sweetgum.nybg.org/ih/

We are grateful for the help of Paul Kirk, CABI, and for the compilers of *Index Fungorum*, whose work made constructing this list much easier. However, we should stress that any errors or omissions are entirely our responsibility.

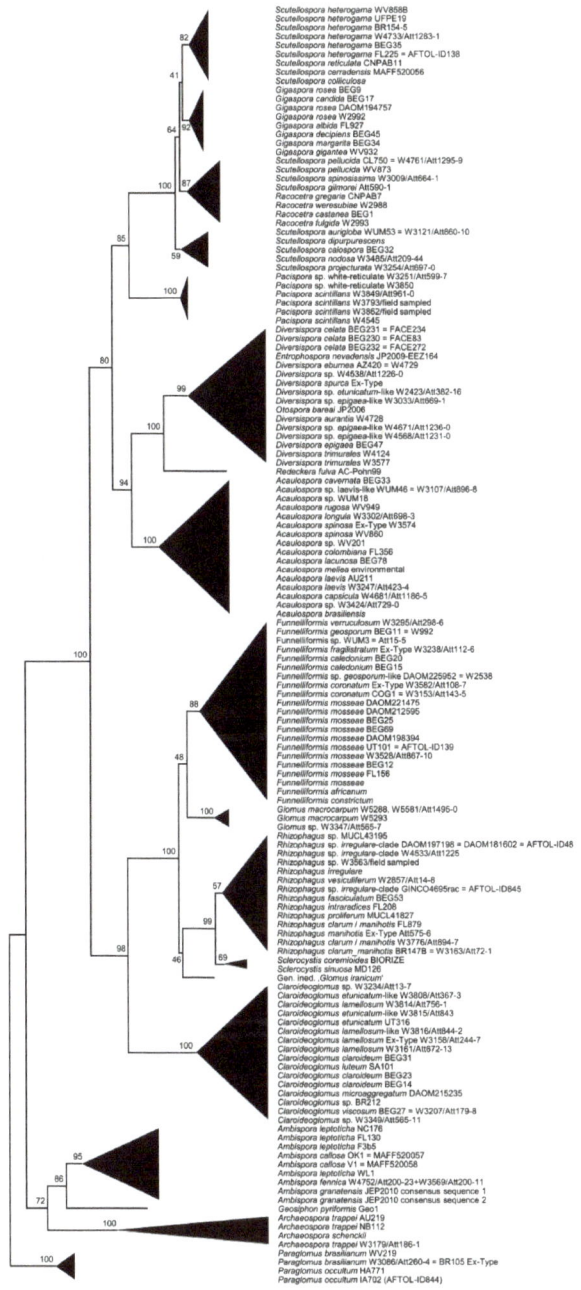

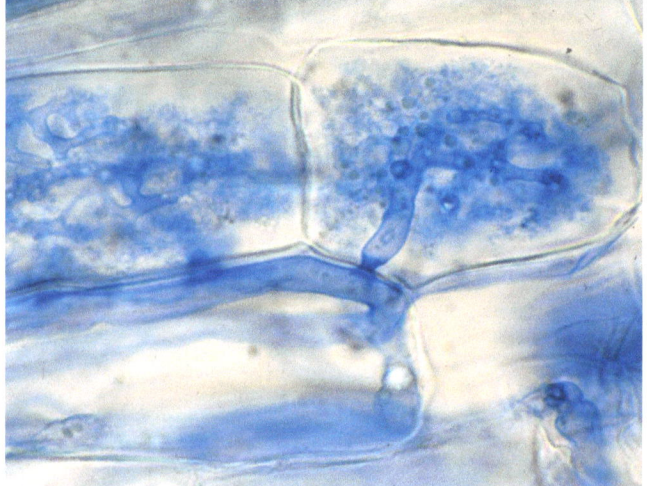

GLOMERALES

Order name: *Glomerales* J.B. Morton & Benny [as *'Glomales'*], Mycotaxon 37: 473 (1990)

GLOMERACEAE

This monogeneric family formerly included all the species in *Glomus* sensu lato, a form-genus that has become a resting place for all of the species with purely glomoid (*Glomus*-like) spores but unknown phylogenetic affiliations. We made suggestions for separation into several genera and two families about a decade ago, but while we were establishing phylogenetic evidence, we refrained from making formal taxonomic separations. We now have the evidence to separate the order into taxa representing a natural classification, based mainly but not exclusively on molecular phylogenetic analyses. The family is now separated into two, *Glomeraceae* based on *Glomus macrocarpum*, and *Claroidoglomeraceae* fam. nov. based on the former *Glomus claroideum*.

Family name: ***Glomeraceae*** Piroz. & Dalpé [as *'Glomaceae'*], Symbiosis 7: 19 (1989) emend C. Walker & A. Schüßler

> Spores glomoid, produced at or near the soil surface, in sporocarps, usually with partial or complete peridium or as spores singly or in clusters in the soil or substrate. With the sequence TGTYADGGCAYYRCACYGG in the ribosomal RNA gene

GLOMUS SENSU STRICTO

This genus is based on the species *Glomus macrocarpum* Tul. & C. Tul. (1845), later lectotypified by Berch & Fortin (1983). For many years, this fungus has been known only from field collections or from unverified pot cultures that have not been available for taxonomic study. We have now established a pot culture and have consequently been able to obtain sequences from the type species for the genus *Glomus*. We also have molecular evidence from two other samples, one from a field collection, and the other from a pot culture, along with morphological comparison with the lectotype of the species. We define an epitype from our culture to allow future comparison. Some of the synonyms indicated by earlier authorities, should be considered as doubtful pending further studies.

Genus name: ***Glomus*** Tul. & C. Tul., G. Bot. Ital. 1 (7-8): 63 (1845) emend C. Walker & A. Schüßler

> ≡*Parapseudoglomus* S.P. Gautam & U.S. Patel, The Mycorrhizae, Diversity, Ecology and Applications (Delhi): 11 (2007)
>
> Spores glomoid, produced at or near the soil surface, in sporocarps, usually with partial or complete peridium or as spores singly or in clusters in the soil or substrate. With the sequences GGTACGYACTGGTATCATTGG and TCGGCTGTAAAAGGCYYTTG in the small subunit ribosomal RNA gene specific for the genus.

Included species:

Generic type: ***Glomus macrocarpum*** Tul. & C. Tul. [as *'macrocarpus'*], G. Bot. Ital. 1(7-8): 63 (1845)

> ≡*Endogone macrocarpa* (Tul. & C. Tul.) Tul. & C. Tul., Fungi Hypog.: 182 (1851)

Specimens examined:

W941 lectotype (see below)

W5288 field collected from which Attempt 1495-0 established from which W5581 was taken for molecular analysis.

Type material

> **Lectotype**. France, Forêt de Chinon, près Ussé, Oct. 1841, Tulasne, determined by Bucholtz, 17 Apr. 1911 to be *Endogone macrocarpa* Tul., No. 13, designated as lectotype by Berch & Fortin (1983), examined 21 Dec 1983 by C. Walker (W941).

> **Epitype**. W5581; 4 Mar. 2009 lodged in the C. Walker collection at **E**, here designated. Derived from the pot culture (Attempt 1495-0); ex-epitype culture material is available on request.

SPECIES OF UNCERTAIN POSITION IN *GLOMUS* SENSU LATO

Glomus achrum Błaszk., D. Redecker, Koegel, Schützek, Oehl & Kovács, Botany 87(3): 262 (2009)

> According to published rDNA sequences, this species clusters basal to the phylogenetic *Glomus* Group Ab and probably belongs to a separate genus that cannot yet be robustly defined.

Glomus aggregatum N.C. Schenck & G.S. Sm., Mycologia 74(1): 80 (1982)

> According to a published β-tubulin sequence analysis, *G. aggregatum* clusters with *Claroideoglomus*, but this seems unlikely and to be potentially eroneous, so we classify it as of uncertain position, for now.

Glomus albidum C. Walker & L.H. Rhodes [as *'albidus'*], Mycotaxon 12(2): 509 (1981)

Glomus ambisporum G.S. Sm. & N.C. Schenck, Mycologia 77(4): 566 (1985)

Glomus antarcticum Cabello, in Cabello, Gaspar & Pollero, Mycotaxon 51: 124 (1994)

Glomus arborense McGee, Trans. Br. Mycol. Soc. 87(1): 123 (1986)

Glomus arenarium Błaszk., Tadych & Madej, Acta Soc. Bot. Pol. 70(2): 97 (2001)

Glomus atrouva McGee & Pattinson, in McGee & Trappe, Aust. Syst. Bot. 15(1): 115 (2002)

Glomus aureum Oehl & Sieverd., in Oehl, Wiemken & Sieverding, J. Appl. Bot., Angew. Bot. 77: 111 (2003)

Glomus australe (Berk.) S.M. Berch, in Berch & Fortin, Can. J. Bot. 61(10): 2611 (1983)

> ≡*Endogone australis* Berk., in Hooker, Flora Tasman., Fungi 2: 282 (1859) [1860]

Glomus avelingiae R.C. Sinclair, in Sinclair, Greuning & Eicker, Mycotaxon 74(2): 338 (2000)

Glomus bagyarajii R.C. Sinclair, in Sinclair, Greuning & Eicker, Mycotaxon 74(2): 338 (2000)

Glomus bistratum Błaszk., D. Redecker, Koegel, Symanczik, Oehl & Kovács, Botany 87(3): 267 (2009)

> According to published rDNA sequences, this species clusters basal to the phylogenetic *Glomus* Group Ab and probably belongs to a separate genus that cannot yet be robustly defined.

Glomus boreale (Thaxt.) Trappe & Gerd. [as *'borealis'*], in Gerdemann & Trappe, Mycol. Mem. 5: 58 (1974)

> ≡*Endogone borealis* Thaxt., Proc. Amer. Acad. Arts & Sci. 57: 318 (1922)

Glomus botryoides F.M. Rothwell & Victor, Mycotaxon 20(1): 163 (1984)

Glomus caesaris Sieverd. & Oehl, in Oehl, Wiemken & Sieverding, Mycotaxon 84: 381 (2002)

Glomus canadense (Thaxt.) Trappe & Gerd. [as *'canadensis'*], in Gerdemann & Trappe, Mycol. Mem. 5: 59 (1974)

> ≡*Endogone canadensis* Thaxt., Proc. Amer. Acad. Arts & Sci. 57: 317 (1922)

Glomus candidum Furrazola, Kaonongbua & Bever, Mycotaxon 113: 103 (2010)

Glomus canum McGee, in McGee & Trappe, Aust. Syst. Bot. 15(1): 116 (2002)

Glomus cerebriforme McGee, Trans. Br. Mycol. Soc. 87(1): 123 (1986)

> According to published rDNA sequences, this species clusters in the phylogenetic *Glomus* Group Ab, but the culture used to generate the sequences may not correspond to the species and is in need of verification, so we here refer to it as of uncertain position.

Glomus citricola D.Z. Tang & M. Zang [as *'citricolum'*], Acta bot. Yunn. 6(3): 301 (1984)

Glomus convolutum Gerd. & Trappe [as *'convolutus'*], Mycol. Mem. 5: 42 (1974)

Glomus corymbiforme Błaszk., Mycologia 87(5): 732 (1995)

Glomus cuneatum McGee & A. Cooper, in McGee & Trappe, Aust. Syst. Bot. 15(1): 117 (2002)

Glomus delhiense Mukerji, Bhattacharjee & J.P. Tewari, Trans. Br. Mycol. Soc. 81(3): 643 (1983)

Glomus deserticola Trappe, Bloss & J.A. Menge, Mycotaxon 20(1): 123 (1984)

> The article in which the molecular evidence for this species is published (Chellappan et al. 2005) shows an illustration of a fungus, but it cannot be verified as *G. deserticola*. The ex-type culture of this species is available, and pending new molecular evidence, we retain this as a species of uncertain phylogenetic position.

Glomus dimorphicum Boyetchko & J.P. Tewari, Can. J. Bot. 64(1): 90 (1986)

Glomus dolichosporum M.Q. Zhang & You S. Wang, in Zhang, Wang & Xing, Mycosystema 16(4): 241 (1997)

Glomus flavisporum (M. Lange & E.M. Lund) Trappe & Gerd. [as *'flavisporus'*], Mycol. Mem. 5: 58 (1974)

Glomus formosanum C.G. Wu & Z.C. Chen, Taiwania 31: 71 (1986)

Glomus fragile (Berk. & Broome) Trappe & Gerd. [as *'fragilis'*], in Gerdemann & Trappe, Mycol. Mem. 5: 59 (1974)

> ≡*Paurocotylis fragilis* Berk. & Broome, J. Linn. Soc., Bot. 14(2): 137 (1875)

Glomus fuegianum (Speg.) Trappe & Gerd. [as *'fuegianus'*], in Gerdemann & Trappe, Mycol. Mem. 5: 58 (1974)

> ≡*Endogone fuegiana* Speg., Anal. Soc. Cient. Argent. 24(3): 125 [no. 5, reprint page 6] (1887)

Glomus gibbosum Błaszk., Mycologia 89(2): 339 (1997)

Glomus globiferum Koske & C. Walker, Mycotaxon 26: 133 (1986)

Glomus glomerulatum Sieverd., Mycotaxon 29: 74 (1987)

Glomus halonatum S.L. Rose & Trappe [as *'halonatus'*], Mycotaxon 10(2): 413 (1980)

Glomus heterosporum G.S. Sm. & N.C. Schenck, Mycologia 77(4): 567 (1985)

Glomus hoi S.M. Berch & Trappe, Mycologia 77(4): 654 (1985)

> Current DNA evidence for organisms that have been determined as *G. hoi* shows that two different cultures fall into two clades separated at the level at least of genus and possibly family. Consequently, we are retaining it in the species of uncertain position pending further clarification.

Glomus hyderabadensis Swarupa, Kunwar, G.S. Prasad & Manohar., Mycotaxon 89(2): 247 (2004)

Glomus indicum Blaszk., Wubet & Harikumar, Botany 88(2): 132-143 (2010)

> According to published rDNA sequences, this species clusters basal to the phylogenetic *Glomus* Group Ab and probably belongs to a separate genus that cannot yet be robustly defined.

Glomus insculptum Błaszk., in Błaszkowski, Adamska & Czerniawska, Mycotaxon 89(2): 227 (2004)

Glomus invermaium I.R. Hall [as *'invermaius'*], Trans. Br. Mycol. Soc. 68(3): 345 (1977)

Glomus kerguelense Dalpé & Strullu, in Dalpé, Plenchette, Frenot, Gloaguen & Strullu, Mycotaxon 84: 53 (2002)

Glomus lacteum S.L. Rose & Trappe [as *'lacteus'*], Mycotaxon 10(2): 415 (1980)

Glomus magnicaule I.R. Hall [as *'magnicaulis'*], Trans. Br. Mycol. Soc. 68(3): 345 (1977)

Glomus melanosporum Gerd. & Trappe [as *'melanosporus'*], Mycol. Mem. 5: 46 (1974)

Glomus microaggregatum Koske, Gemma & P.D. Olexia, Mycotaxon 26: 125 (1986)

Glomus microcarpum Tul. & C. Tul. [as *'microcarpus'*], G. Bot. Ital. 1(7-8): 63 (1845)

Glomus minutum Błaszk., Tadych & Madej, Mycotaxon 76: 189 (2000)

Glomus monosporum Gerd. & Trappe [as *'monosporus'*], Mycol. Mem. 5: 41 (1974)

> This has always been something of a problem. It was established in pot culture, but the culture was lost long before DNA sequencing methods were applied as markers for clades in the *Glomeromycota*. There have been several cultured organisms that have been given this name, but none seems properly to fit the species description of brown spores that have ornamentation of spines on the laminated wall component.

Glomus mortonii Bentiv. & Hetrick, Mycotaxon 42: 10 (1991)

> Although this fungus has produced spores in soil traps in Finland, it has not yet been possible either to establish it in pure culture or to extract DNA. Consequently, it is unclear as to whether or not it belongs in any of the established clades.

Glomus multicaule Gerd. & B.K. Bakshi [as *'multicaulis'*], Trans. Br. Mycol. Soc. 66(2): 340 (1976)

Glomus multiforum Tadych & Błaszk., in Blaszkowski & Tadych, Mycologia 89(5): 805 (1997)

Glomus nanolumen Koske & Gemma, Mycologia 81(6): 935 (1990) [1989]

Glomus pallidum I.R. Hall [as *'pallidus'*], Trans. Br. Mycol. Soc. 68(3): 343 (1977)

Glomus pansihalos S.M. Berch & Koske, Mycologia 78(5): 832 (1986)

Glomus pellucidum McGee & Pattinson, in McGee & Trappe, Aust. Syst. Bot. 15(1): 120 (2002)

Glomus perpusillum Błaszk. & Kovács, in Błaszkowski, Kovács & Balázs, Mycologia 101(2): 249 (2009)

Glomus przelewicense Błaszk. [as *'przelewicensis'*], Bulletin of the Polish Academy of Sciences, Biological Sciences 36(10-12): 272 (1988)

Glomus pulvinatum (Henn.) Trappe & Gerd. [as *'pulvinatus'*], in Gerdemann & Trappe, Mycol. Mem. 5: 59 (1974)

> ≡*Endogone pulvinata* Henn., Hedwigia 36: 212 (1897)

Glomus pustulatum Koske, Friese, C. Walker & Dalpé, Mycotaxon 26: 143 (1986)

Glomus radiatum (Thaxt.) Trappe & Gerd. [as *'radiatus'*], in Gerdemann & Trappe, Mycol. Mem. 5: 46 (1974)

> ≡*Endogone radiata* Thaxt., Proc. Amer. Acad. Arts & Sci. 57: 316 (1922)

Glomus reticulatum Bhattacharjee & Mukerji [as *'reticulatus'*], Sydowia 33: 14 (1980)

> This has such a vague and inadequately illustrated species description that it is impossible to apply the name with any confidence to any organism. The type material seems not to exist, and it is thus a species name that exists but which cannot be usefully applied.

Glomus segmentatum Trappe, Spooner & Ivory [as *'segmentatus'*], in Trappe, Trans. Br. Mycol. Soc. 73(2): 362 (1979)

> This is a truly sporocarpic fungus. It is known only from field collections, but no sequence exists that can be used to establish its true phylogeny.

Glomus spinosum H.T. Hu, Mycotaxon 83: 160 (2002)

> The species description is more or less uninterpretable, and no type material exists for any of Hu's species.

Glomus spinuliferum Sieverd. & Oehl, in Oehl, Wiemken & Sieverding, Mycotaxon 86: 158 (2003)

Glomus sterilum V.S. Mehrotra & Baijal [as *'sterile'*], Philipp. J. Sci., C, Bot. 121(3): 306 (1992)

> This is a *nom. inval.* (invalid name), having been published without a Latin description or diagnosis (a requirement of the Botanical Code for formal descriptions after 1 Jan 1935.

Glomus tenebrosum (Thaxt.) S.M. Berch, in Berch & Fortin, Can. J. Bot. 61(10): 2615 (1983)

> ≡*Endogone tenebrosa* Thaxt., Proc. Amer. Acad. Arts & Sci. 57: 314 (1922)

Glomus tenerum P.A. Tandy [as *'tener'*], Aust. J. Bot. 23(5): 864 (1975)

Glomus tenue (Greenall) I.R. Hall [as *'tenuis'*], Trans. Br. Mycol. Soc. 68(3): 350 (1977)

> ≡*Rhizophagus tenuis* Greenall, N.Z. J. Bot. 1(4): 398 (1963)

> A fungus was published under the name *Rhizophagus tenuis* and later transferred to the genus *Glomus*. As used, the name is unlikely to refer to either a single fungus, or a member of *Glomus*. Fungi fitting the description have from time to time been established in pot culture, but it seems none has ever been purified and there is to date no molecular data with which to place it in its correct clade. It has almost no similarities with any other fungus in the phylum, even more so now that it is known that glomoid spores have been shown to occur in many widely separated taxa.

Glomus tortuosum N.C. Schenck & G.S. Sm., Mycologia 74(1): 83 (1982)

Glomus versiforme (P. Karst.) S.M. Berch, in Berch & Fortin, Can. J. Bot. 61(10): 2614 (1983).

> ≡*Endogone versiformis* P. Karst., Hedwigia 23: 39 (1884), non sensu Berch & Fortin, Can. J. Bot. 61: 2614 (1983).

> This species is now confounded by its erroneous synonymisation with the former *Glomus epigaeum*. We have evidence to be published elsewhere showing that these two species are not synonymous, and therefore the species is re-defined through a lectotype.

> Specimens examined: Finland, Nylandia, Helsinfors (Helsinki); from the substratum in a plant pot in a cold glasshouse, *Cercocarpus ledifolia*, '23. XI. 1860 - I. 1861' [*sic*], leg. W. Nylander, Mus. Bot. Univ., Helsinki 3936 p.p., lectotype W4551 in **H**, here designated; isolectotype in **E**.

Glomus viscosum T.H. Nicolson, in Walker, Giovannetti, Avio, Citernesi & Nicolson, Mycol. Res. 99(12): 1502 (1995)

> Although this fungus has ostensibly been placed next to *G. claroideum* from sequence data, it seems the culture from which the evidence came also contained spores of other AM fungi. Consequently, we wait to verify its position from a pure culture before placing it in any new genus.

Glomus warcupii McGee, Trans. Br. Mycol. Soc. 87(1): 125 (1986)

Glomus zaozhuangianus F.Y. Wang & R.J. Liu [as *'zaozhuangianus'*], Mycosystema 21(4): 522 (2002)

The following were synonymised with *G. macrocarpum* by Gerdemann & Trappe (1974). Because we have no molecular evidence, we have moved them to the list of *Glomus* sensu lato of uncertain phylogenetic position.

Endogone australis Berk., in Hooker, Flora Tasman., Fungi 2: 282 (1859) [1860]

Paurocotylis fulva var. *zealandica* Cooke [as *'zælandica'*], Grevillea 8 (no. 46): 59 (1879)

Endogone pampaloniana Bacc., G. Bot. Ital.,Part 2. 10: 79 (1903)

Endogone nuda Petch, Ann. R. Bot. Gdns Peradeniya 9: 322 (1925)

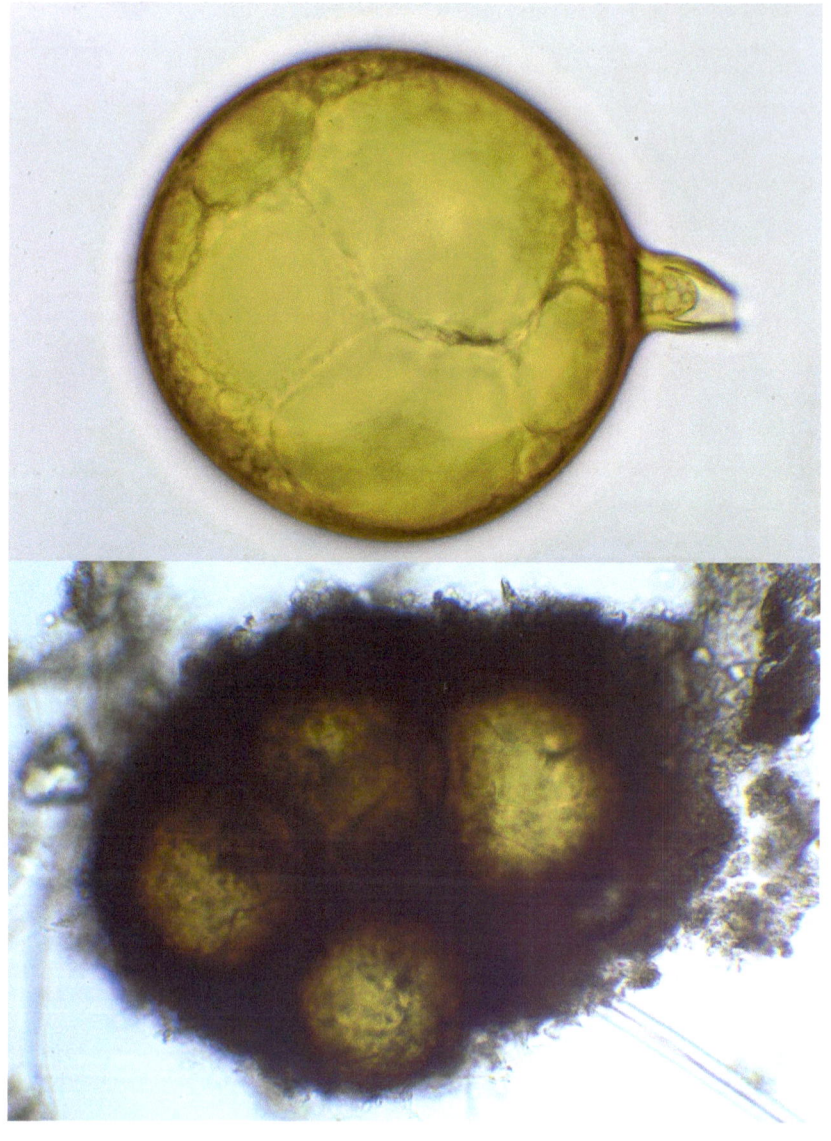

FUNNELIFORMIS

It has been clear for many years that the fungus, *Glomus mosseae* has characteristics quite different from those of *Glomus macrocarpum*. Now that the generic type of *Glomus* has been cultured and sequenced, the evidence is available to separate this fungus and other related organisms from that genus and place them in a newly named taxon at the generic level.

Funneliformis C. Walker & A. Schüßler gen. nov.

Latin diagnosis:

> A generibus ceteris in Glomeromycota combinatus sporophoro glomoideo et sequentio DNA differenti. Mycorrhizas arbusculares formans. Sequentia typica acidi desoxyribonucleici monadis 'SSU' ribosomatum: CGGTCATGCCGTTGGTATGY.

> Differs from other glomoid spore producing genera in the *Glomeromycota* by coloured spores formed in the soil or substratum in sporocarps of 1 to approx. 20 spores surrounded by an entire or partial coarse mycelial mantle, or ectocarpic spores singly or in loose clusters in the substratum. Spores often with a funnel-shaped spore base. Spore wall structure normally of two or three components in a single wall group. Outer component colourless, often disappearing as the spore matures. Spores normally occluded by a septum in the subtending hypha distal to the spore base. With the sequence CGGTCATGCCGTTGGTATGY of the small subunit ribosomal RNA gene specific for the genus. Forming arbuscular mycorrhizas.

Typus: *Endogone mosseae* T.H. Nicolson & Gerd. (1968)

Type material: holotype – Scotland, Perthshire, Benvie, October 1961 **FH**

> **Epitype**: W5790; 23 Jun 2010 in the C. Walker collection in **E**, here designated. Derived from the culture with the designator Attempt 109-28 that had Attempt 109-20, a single spore isolate of C. Walker that originated from Attempt 109-0 (offspring of which became BEG12), a multi-spore culture from East Malling, known as the 'Rothamsted *G. mosseae*' in its ancestry); ex-epitype culture material is available on request and will be donated to at least one international collection.

> **Etymology**: *Funneliformis* referring the often funnel-shaped spore base of species in the genus.

Included species

Generic type: ***Funneliformis mosseae*** (T.H. Nicolson & Gerd.) C. Walker & A. Schüßler comb. nov.

> ≡Endogone mosseae T.H. Nicolson & Gerd., Mycologia 60(2): 314 (1968)

> ≡Glomus mosseae (T.H. Nicolson & Gerd.) Gerd. & Trappe, Mycol. Mem. 5: 40 (1974)

Funneliformis caledonium (T.H. Nicolson & Gerd.) C. Walker & A. Schüßler comb. nov.

> ≡*Endogone macrocarpa* var. *caledonia* T.H. Nicolson & Gerd., Mycologia 60(2): 322 (1968)

> ≡*Glomus caledonium* (T.H. Nicolson & Gerd.) Trappe & Gerd. [as '*caledonius*'], in Gerdemann & Trappe, Mycol. Mem. 5: 56 (1974)

Funneliformis badium (Oehl, D. Redecker & Sieverd.) C. Walker & A. Schüßler comb. nov.

> ≡*Glomus badium* Oehl, D. Redecker & Sieverd., J Appl Bot Food Qual 79: 39 (2005)

Funneliformis africanum (Błaszk. & Kovács) C. Walker & A. Schüßler comb. nov.

> ≡*Glomus africanum* Błaszk. & Kovács Mycologia 102(6): 1452 (2010)

Funneliformis coronatum (Giovann.) C. Walker & A. Schüßler comb. nov.

> ≡*Glomus coronatum* Giovann., in Giovannetti, Avio & Salutini, Can. J. Bot. 69(1): 162 (1991)

Funneliformis fragilistratum (Skou & I. Jakobsen) C. Walker & A. Schüßler comb. nov.

> ≡*Glomus fragilistratum* Skou & I. Jakobsen, Mycotaxon 36(1): 276 (1989)

Funneliformis geosporum (T.H. Nicolson & Gerd.) C. Walker & A. Schüßler comb. nov.

≡*Endogone macrocarpa* var. *geospora* T.H. Nicolson & Gerd., Mycologia 60(2): 318 (1968)

≡*Glomus geosporum* (T.H. Nicolson & Gerd.) C. Walker, Mycotaxon 15: 56 (1982)

Funneliformis verruculosum (Błaszk.) C. Walker & A. Schüßler comb. nov.

≡*Glomus verruculosum* Błaszk., in Blaszkowski & Tadych, Mycologia 89(5): 809 (1997)

Funneliformis vesiculiferum (Thaxt.) C. Walker & A. Schüßler comb. nov.

≡*Endogone vesiculifera* Thaxt., Proc. Amer. Acad. Arts & Sci. 57: 309 (1922)

≡*Glomus vesiculiferum* (Thaxt.) Gerd. & Trappe [as *'vesiculifer'*], Mycol. Mem. 5: 49 (1974)

Funneliformis constrictum (Trappe) C. Walker & A. Schüßler comb. nov.

≡*Glomus constrictum* Trappe [as *'constrictus'*], Mycotaxon 6(2): 361 (1977)

Funneliformis xanthium (Błaszk., Blanke, Renker & Buscot) C. Walker & A. Schüßler comb. nov.

≡*Glomus xanthium* Błaszk., Blanke, Renker & Buscot, Mycotaxon 90(2): 459 (2004)

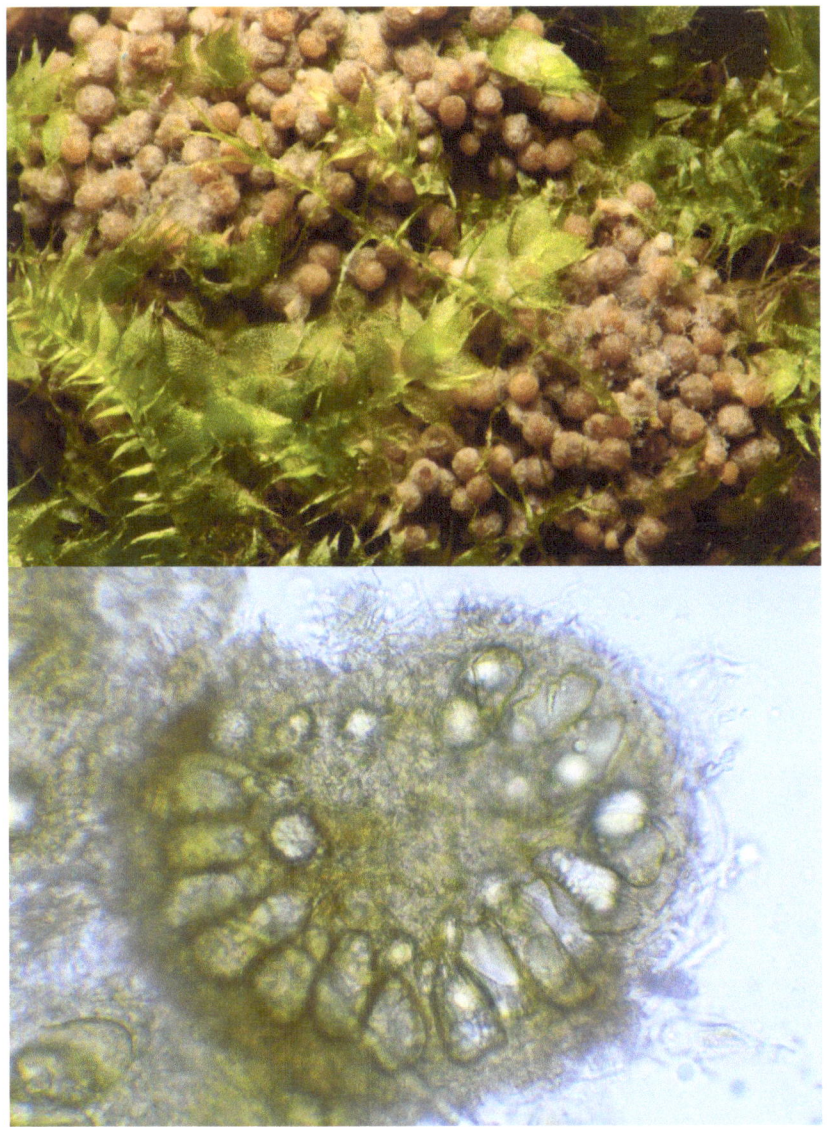

SCLEROCYSTIS

This genus was described in 1875 (Berkeley & Broome 1873), but was united with *Glomus* (Almeida & Schenck 1990). We now separate it once more based on the published molecular evidence for two species (Redecker et al. 2000) and the basal phylogenetic placement basal to *Glomus* Group Ab (*Rhizophagus*). More evidence will be required before anything further can be concluded as to the natural phylogenetic position of the remaining species.

Genus name: ***Sclerocystis*** Berk. & Broome, J. Linn. Soc., Bot. 14(2): 137 (1875)

> Forming glomoid spores in sporocarps with a peridium, radiating from a central sterile plexus of mycelium.

Included species

Generic type: ***Sclerocystis coremioides*** Berk. & Broome, J. Linn. Soc., Bot. 14(2): 137 (1875)

> ≡*Glomus coremioides* (Berk. & Broome) D. Redecker & J.B. Morton, in Redecker, Morton & Bruns 2000

Sclerocystis sinuosa Gerd. & B.K. Bakshi, Trans. Br. Mycol. Soc. 66(2): 343 (1976)

> ≡*Glomus sinuosum* (Gerd. & B.K. Bakshi) R.T. Almeida & N.C. Schenck, Mycologia 82(6): 710 (1990)

> =Sclerocystis *pakistanica* S.H. Iqbal & Perveen, Trans. Mycol. Soc. Japan 21(1): 59 (1980)

SPECIES OF UNCERTAIN POSITION IN *SCLEROCYSTIS*

Sclerocystis alba Petch, Ann. R. Bot. Gdns Peradeniya 9: 322 (1925)

> ≡*Endogone alba* (Petch) Gerd. & Trappe, Mycol. Mem. 5: 25 (1974)

>> This species was transferred to *Endogone* by Gerdemann & Trappe (1974), but from their description, it may be worthwile to reconsider its position should it ever be possible to obtain molecular evidence.

Sclerocystis dussii (Pat.) Höhn., Sber. Akad. Wiss. Wien, Math.-Naturw. Kl., Abt. 1 119: 399 [7 repr.] (1910)

> ≡*Ackermannia dussii* Pat., Bull. Soc. Mycol. Fr. 18(2): 181 (1902)

> ≡*Sphaerocreas dussii* (Pat.) Höhn., Sber. Akad. Wiss. Wien, Math.-Naturw. Kl., Abt. 1 118: 401 [127 repr.] (1909)

Sclerocystis coccogenum (Pat.) Höhn., Sber. Akad. Wiss. Wien, Math.-Naturw. Kl., Abt. 1 119: 399 [7 repr.] (1910)

> ≡*Ackermannia coccogena* Pat., Bull. Soc. Mycol. Fr. 18(2): 182 (1902)

> ≡*Sphaerocreas coccogenum* (Pat.) Höhn., Sber. Akad. Wiss. Wien, Math.-Naturw. Kl., Abt. 1 118: 401 [127 repr.] (1909)

Sclerocystis microcarpus S.H. Iqbal & Perveen, Trans. Mycol. Soc. Japan 21: 58 (1980)

Sclerocystis liquidambaris C.G. Wu & Z.C. Chen, Trans. Mycol. Soc. Rep. China 2(2): 74 (1987)

> =*Glomus liquidambaris* (C.G. Wu & Z.C. Chen) R.T. Almeida & N.C. Schenck, Mycologia 82(6): 711 (1990)

> =*Glomus liquidambaris* (C.G. Wu & Z.C. Chen) R.T. Almeida & N.C. Schenck ex Y.J. Yao, in Yao, Pegler & Young, Kew Bull. 50(2): 306 (1995)

> =*Sclerocystis cunninghamia* H.T. Hu, Quarterly Journal of Chinese Forestry 21(2): 52 (1988)

Sclerocystis rubiformis Gerd. & Trappe, Mycol. Mem. 5: 60 (1974)

> ≡*Glomus rubiforme* (Gerd. & Trappe) R.T. Almeida & N.C. Schenck, Mycologia 82(6): 709 (1990)

> =*Sclerocystis indica* Bhattacharjee & Mukerji, in Bhattacharjee, Mukerji & Misra, Acta Bot. Indica 8(1): 99 (1980)

> =*Sclerocystis pachycaulis* C.G. Wu & Z.C. Chen, Taiwania 31: 74 (1986)

Sclerocystis pubescens (Sacc. & Ellis) Höhn., Sber. Akad. Wiss. Wien, Math.-Naturw. Kl., Abt. 1 119: 399 [7 repr.] (1910)

≡*Sphaerocreas pubescens* Sacc. & Ellis, Michelia 2 (no. 8): 582 (1882)

≡*Endogone pubescens* (Sacc. & Ellis) Zycha, Krypt.-Fl. Brandenburg (Leipzig) 6a(2): 214 (1935)

≡*Stigmatella pubescens* (Sacc. & Ellis) Sacc., Syll. Fung. (Abellini) 4: 680 (1886)

≡*Glomus pubescens* (Sacc. & Ellis) Trappe & Gerd., in Gerdemann & Trappe, Mycol. Mem. 5: 57 (1974)

Sclerocystis taiwanensis C.G. Wu & Z.C. Chen, Trans. Mycol. Soc. Rep. China 2(2): 78 (1987)

≡*Glomus taiwanense* (C.G. Wu & Z.C. Chen) R.T. Almeida & N.C. Schenck, Mycologia 82(6): 711 (1990)

≡*Glomus taiwanense* (C.G. Wu & Z.C. Chen) R.T. Almeida & N.C. Schenck ex Y.J. Yao, in Yao, Pegler & Young, Kew Bull. 50(2): 306 (1995)

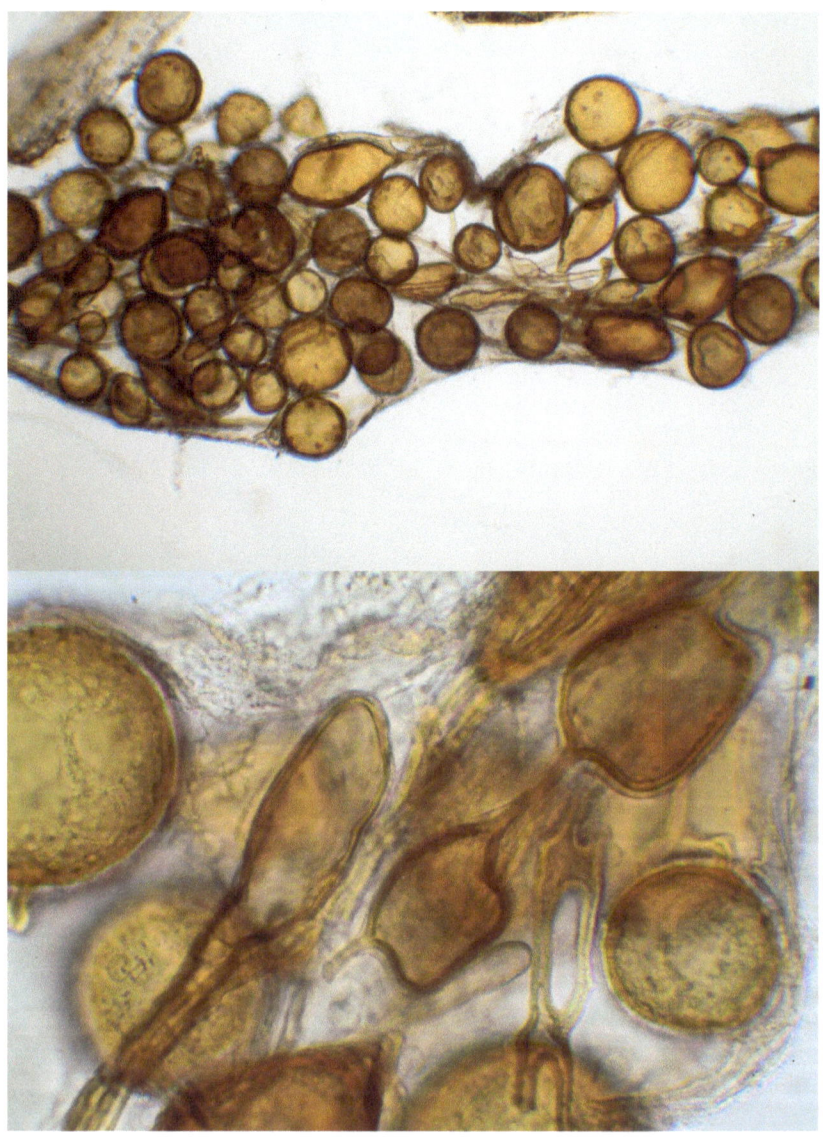

RHIZOPHAGUS

Gerdemann & Trappe (1974) synonymised *Rhizophagus* with *Glomus*. However, examination of the protologue of *R. populinus* reveals that the species is an arbuscular mycorrhizal fungus that produces abundant spores in roots. This is a characteristic of the species in the phylogenetic group *Glomus* Group Ab (GlGrAb), and this group falls into a clade at the generic level. Consequently, we resurrect the genus *Rhizophagus* for species in the *Glomeraceae* that form abundant spores in the roots of vascular plants.

Genus name: *Rhizophagus* P.A. Dang., Botaniste 5: 43 (1896)

Included species

Generic type: *Rhizophagus populinus* P.A. Dang., Botaniste 5: 285 (1896)

> Glomoid spores formed singly, in loose clusters, tight clusters (fascicles) in the substrate and in the roots or rhizoids of host plants. Forming arbuscular mycorrhizas.

Rhizophagus clarus (T.H. Nicolson & N.C. Schenck) C. Walker & A. Schüßler comb. nov.

> ≡*Glomus clarum* T.H. Nicolson & N.C. Schenck [as *'clarus'*], Mycologia 71(1): 182 (1979)

Rhizophagus diaphanus (C. Cano & Y. Dalpé) C. Walker & A. Schüßler comb. nov.

> ≡*Glomus diaphanum* J.B. Morton & C. Walker, Mycotaxon 21: 433 (1984)

Rhizophagus custos (C. Cano & Y. Dalpé) C. Walker & A. Schüßler comb. nov.

> ≡*Glomus custos* C. Cano & Dalpé, Mycotaxon 109: 502 (2009)

Rhizophagus fasciculatus (Thaxt.) C. Walker & A. Schüßler comb. nov.

> ≡*Endogone fasciculata* Thaxt., Proc. Amer. Acad. Arts & Sci. 57: 308 (1922)

> ≡*Glomus fasciculatum* (Thaxt.) Gerd. & Trappe [as *'fasciculatus'*], Mycol. Mem. 5: 51 (1974)

Rhizophagus intraradices (N.C. Schenck & G.S. Sm.) C. Walker & A. Schüßler comb. nov.

> ≡*Glomus intraradices* N.C. Schenck & G.S. Sm., Mycologia 74(1): 78 (1982)

Rhizophagus iranicus (Błaszk., Kovács & Balázs) C. Walker & A. Schüßler comb. nov.

> ≡*Glomus iranicum* Błaszk. & Kovács Mycologia 102(6): 1457 (2010)

Rhizophagus irregulare (Błaszk., Wubet, Renker & Buscot) C. Walker & A. Schüßler comb. nov.

> ≡*Glomus irregulare* Błaszk., Wubet, Renker & Buscot, Mycotaxon 106: 252 (2008)

Rhizophagus manihotis (R.H. Howeler, Sieverd. & N.C. Schenck) C. Walker & A. Schüßler comb. nov.

> ≡*Glomus manihotis* R.H. Howeler, Sieverd. & N.C. Schenck, in Schenck, Spain, Sieverding & Howeler, Mycologia 76(4): 695 (1984)

Rhizophagus proliferus (Błaszk., Kovács & Balázs) C. Walker & A. Schüßler comb. nov.

> *Glomus proliferum* Dalpé & Declerck, in Declerck, Cranenbrouck, Dalpé, Séguin, Grandmougin-Ferjani, Fontaine & Sancholle, Mycologia 92(6): 1180 (2000)

CLAROIDEOGLOMERACEAE

This group is the phylogenetic *Glomus* Group B (GlGrB; Schüßler et al. 2001a), separating well in phylogenetic trees at the level of a family (Schwarzott et al. 2001).

Family name: *Claroideoglomeraceae* C. Walker & A. Schüßler fam. nov.

Latin diagnosis

> A familiis ceteris in Glomeromycota combinatus sporophoro glomoideo et sequentio DNA differenti. Mycorrhizas arbusculares formans. Sequentia typica acidi desoxyribonucleici monadis 'SSU' ribosomatum: CAGYYGGGRAACCRACTAAA; ATTKRCACATCGGTCGTGCCYTAAGGGGYATGAACYRGTGTAGTSA; TAAAA YRGGACGGCATGATTCTATT.

> Producing glomoid spores that form in the substrate or rarely as individual spores in decaying roots and with a combination of an evanescent outer wall component and an inner semi-flexible component that forms an apparent 'endospore' at maturity and having the sequence CAGYYGGGRAACCRACTAAA; ATTKRCACATCGGTCGTGC CYTAAGGGGYATGAACYRGTGTAGTSA; TAAAAYRGGACGGCATGATTCTATT of the small subunit ribosomal RNA gene specific for the family. Forming arbuscular mycorrhizas.

> **Typus** *Claroideoglomus* C. Walker & A. Schüßler gen. nov.

CLAROIDEOGLOMUS

This is presently the only genus in the family *Claroideoglomeraceae*. Like the family, it corresponds to the phylogenetic *Glomus* Group B. All spores have an inner wall developed separately.

Claroideoglomus C. Walker & A. Schüßler gen. nov.

Latin diagnosis

> A generibus ceteris in Glomeromycota combinatus sporophoro glomoideo et sequentio DNA differenti. Mycorrhizas arbusculares formans. Sequentia typica acidi desoxyribonucleici monadis 'SSU' ribosomatum: CAGYTGGGRAACCRACTAAA; ATTKRCACATCGGTCGTGCC.

> Differs from other genera in the *Glomeromycota* by possession of glomoid spores that form in the substrate or rarely as individual spores in decaying roots and with a combination of an evanescent outer wall component and an inner semi-flexible component that seemingly forms an 'endospore' at maturity. With the sequence CAGYTGGGRAACCRACTAAA; ATTKRCACATCGGTCGTGCC of the small subunit ribosomal RNA gene specific for the genus. Forming arbuscular mycorrhizas.

> **Typus**: *Glomus claroideum* N.C. Schenck & G.S. Sm. (1982).

> **Type material**: USA, Florida, Sanford, Agricultural Research and Education Center, July 1976, **OSC** 40252, Isotypes **FH**, **FLAS** F52578.

> **Etymology**: *Claroideoglomus,* referring to the glomoid spores sharing the developmental pattern and wall structure of *Glomus claroideum.*

Included species

Generic type: *Claroideoglomus claroideum* (N. C. Schenck & G. S. Sm.) C. Walker & A. Schüßler comb. nov.

> ≡*Glomus claroideum* N.C. Schenck & G.S. Sm., Mycologia 74(1): 84 (1982)

> =*Glomus maculosum* D.D. Mill. & C. Walker, Mycotaxon 25(1): 218 (1986)

> =*Glomus fistulosum* Skou & I. Jakobsen, Mycotaxon 36(1): 274 (1989)

> **Epitype**: The epitype, W2370 from Attempt 564 was designated by Walker and Vestberg 1998 and is available as ex-type material on request.

Claroideoglomus walkeri (Błaszk. & C. Renker) C. Walker & A. Schüßler comb. nov.

≡*Glomus walkeri* Błaszk. & Renker, in Błaszkowski, Renker & Buscot, Mycol. Res. 110(5): 563 (2006)

Claroideoglomus luteum (L.J. Kenn., J.C. Stutz & J.B. Morton) C. Walker & A. Schüßler comb. nov.

≡*Glomus luteum* L.J. Kenn., J.C. Stutz & J.B. Morton, Mycologia 91(6): 1090 (1999)

Claroideoglomus lamellosum (Dalpé, Koske & Tews) C. Walker & A. Schüßler comb. nov.

≡*Glomus lamellosum* Dalpé, Koske & Tews, Mycotaxon 43: 289 (1992)

Claroideoglomus etunicatum (W.N. Becker & Gerd.) C. Walker & A. Schüßler comb. nov.

≡*Glomus etunicatum* W.N. Becker & Gerd. [as *'etunicatus'*], Mycotaxon 6(1): 29 (1977)

Claroideoglomus drummondii (Błaszk. & C. Renker) C. Walker & A. Schüßler comb. nov.

≡*Glomus drummondii* Błaszk. & Renker, in Błaszkowski, Renker & Buscot, Mycol. Res. 110(5): 559 (2006)

SPECIES OF UNCERTAIN POSITION IN *CLAROIDEOGLOMUS*

The following species name has at some time been synonymised with members that now occupy the above genus, but the species cannot be verified from molecular evidence, and thus has now been moved into uncertain positions.

Glomus multisubstensum Mukerji, Bhattacharjee & J.P. Tewari, Trans. Br. Mycol. Soc. 81(3): 641 (1983)

SPECIES OF UNCERTAIN POSITION IN *GLOMERALES*

The following species were synonymised from morphological evidence (Gerdemann & Trappe 1974), but because there is no supporting molecular data, we have left them as species of uncertain position.

Formerly synonymised with *Glomus fulvum*:

Endogone lignicola Pat., Bull. Soc. Mycol. Fr. 18(2): 183 (1902)

Endogone moelleri Henn., Hedwigia 36: 211 (1897)

Formerly synonymised with *Glomus fasciculatum*:

Endogone arenacea Thaxt., Proc. Amer. Acad. Arts & Sci. 57: 317 (1922)

Rhizophagites butleri Rosend., Bull. Torrey Bot. Club 70: 131 (1943)

The following was synonymised with *G. fasciculatum*, but not by Gerdemann & Trappe.

Palaeomycites butleri (F. Rosend.) Kalgutkar & Janson., AASP Contributions Series (Dallas) 39: 208 (2000)

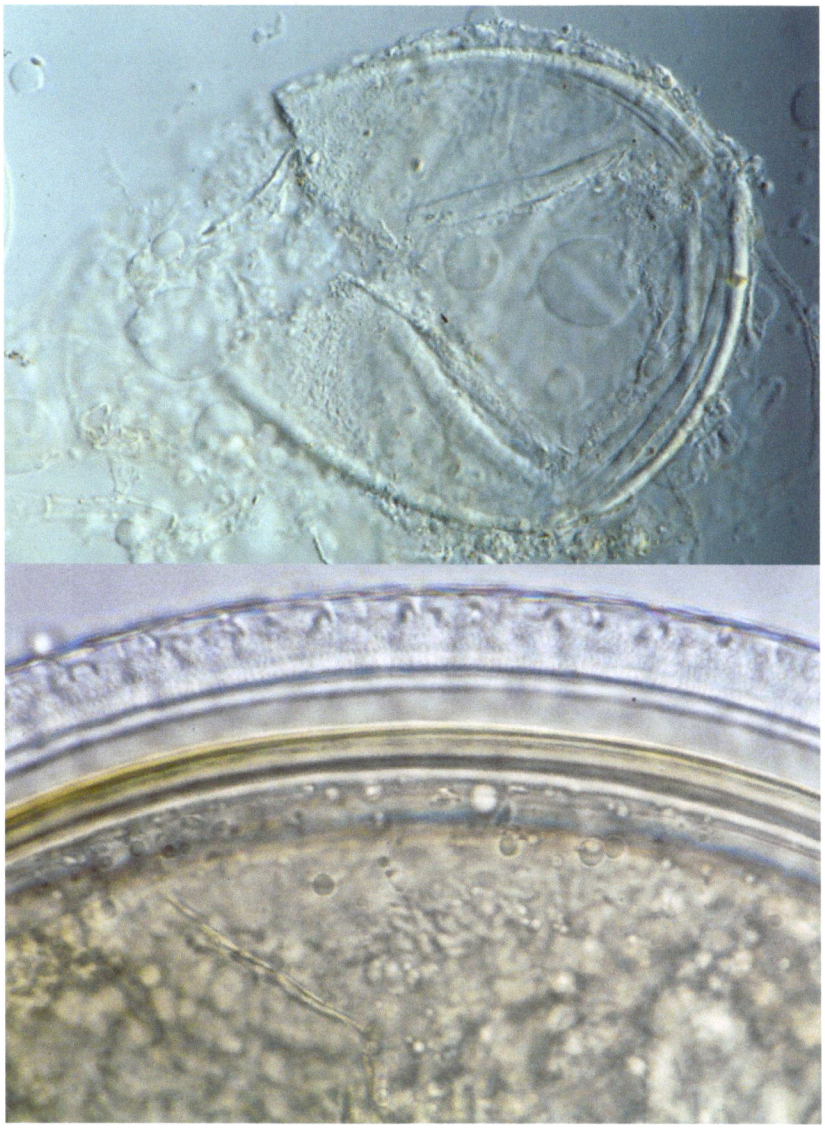

DIVERSISPORALES

Order name: *Diversisporales* C. Walker & A. Schüßler, Mycol. Res. 108(9): 981 (2004)

GIGASPORACEAE

Family name: *Gigasporaceae* J.B. Morton & Benny, Mycotaxon 37: 483 (1990)

Although there is clear indication that one additional genus will be required to apply a natural systematics concept in this family, its generic structure currently is retained as per Morton and Msiska (2010) following their suggestions to await future robust phylogenetic evidence.

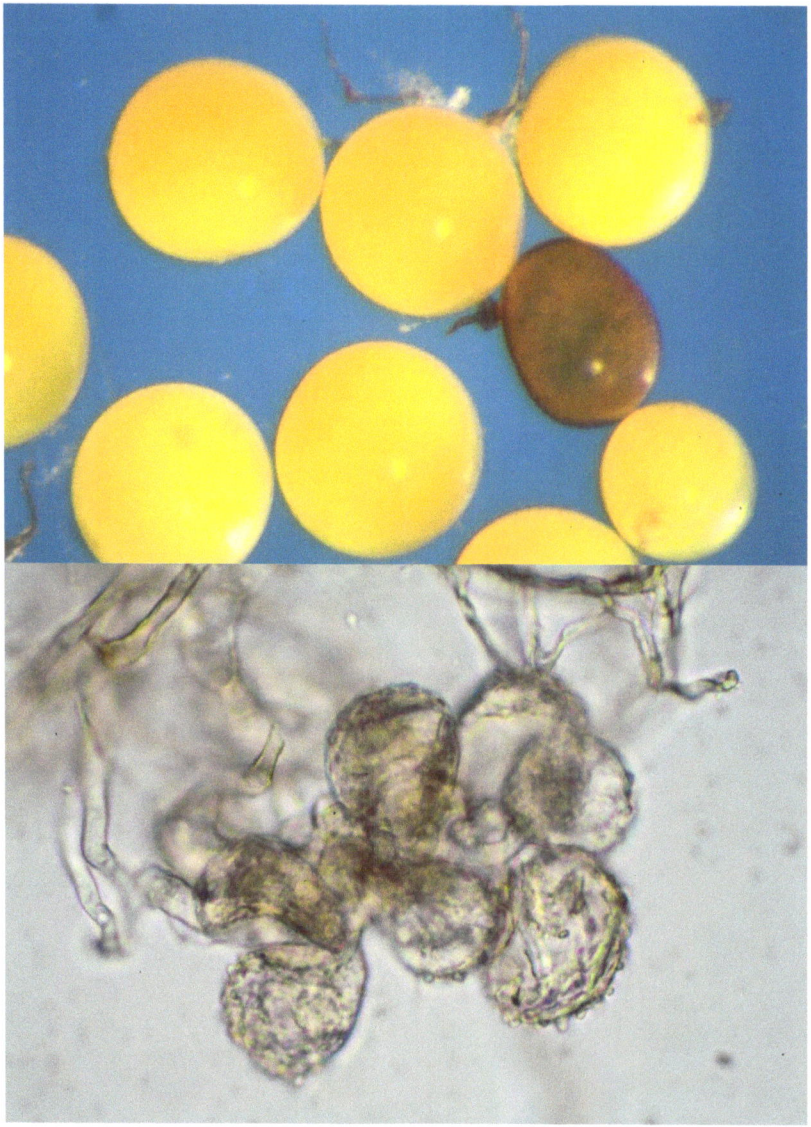

GIGASPORA

Genus name: *Gigaspora* Gerd. & Trappe, Mycol. Mem. 5: 25 (1974)

Included species

Generic type: *Gigaspora gigantea* (T.H. Nicolson & Gerd.) Gerd. & Trappe, Mycol. Mem. 5: 29 (1974)

> ≡*Endogone gigantea* T.H. Nicolson & Gerd., Mycologia 60(2): 321 (1968)

Gigaspora albida N.C. Schenck & G.S. Sm., Mycologia 74(1): 85 (1982)

Gigaspora candida Bhattacharjee, Mukerji, J.P. Tewari & Skoropad, Trans. Br. Mycol. Soc. 78(1): 184 (1982)

> Although the type material of *G. candida* is lost, the species description is such that it cannot be separated from *G. alboaurantiaca*. We have therefore placed the latter as a later heterotypic synonym of the former.

> **Type**: Delhi, Wasirabad, March 1980 DU/KMB 494, **DUH**.

>> = *Gigaspora alboaurantiaca* W.N. Chou, in Chou, Yen & Chung, Trans. Mycol. Soc. Rep. China 6(3-4): 3 (1991)

>> **Type**: from one of the following places: Taiwan-Taipei, Chinshan, Taoyuan at Haihu, Tainan at Anpien, or Hualien at Hoping; from a pot culture with Zea mays made from rhizosphere of Casuarina equisetifolia, 25 Jul 1987, Chou-2010 **TFRI**

> **Epitype**: 5136, 5 Mar 2007, in the C. Walker collection in **E**, here designated. Derived from the culture with the designator Attempt 26-32 that had Attempt 26-0, a single spore isolate of W. N. Chou, in its ancestry; the ex-epitype culture material (also the ex-type culture of *Gigaspora alboaurantiaca*) is available from the BEG collection (www.kent.ac.uk/bio/beg/englishhomepage.htm) as BEG17.

Gigaspora decipiens I.R. Hall & L.K. Abbott, Trans. Br. Mycol. Soc. 83(2): 204 (1984)

Gigaspora gigantea (T.H. Nicolson & Gerd.) Gerd. & Trappe, Mycol. Mem. 5: 29 (1974)

> ≡*Endogone gigantea* T.H. Nicolson & Gerd., Mycologia 60(2): 321 (1968)

Gigaspora margarita W.N. Becker & I.R. Hall, Mycotaxon 4(1): 155 (1976)

Gigaspora ramisporophora Spain, Sieverd. & N.C. Schenck, Mycotaxon 34(2): 668 (1989)

Gigaspora rosea T.H. Nicolson & N.C. Schenck, Mycologia 71(1): 190 (1979)

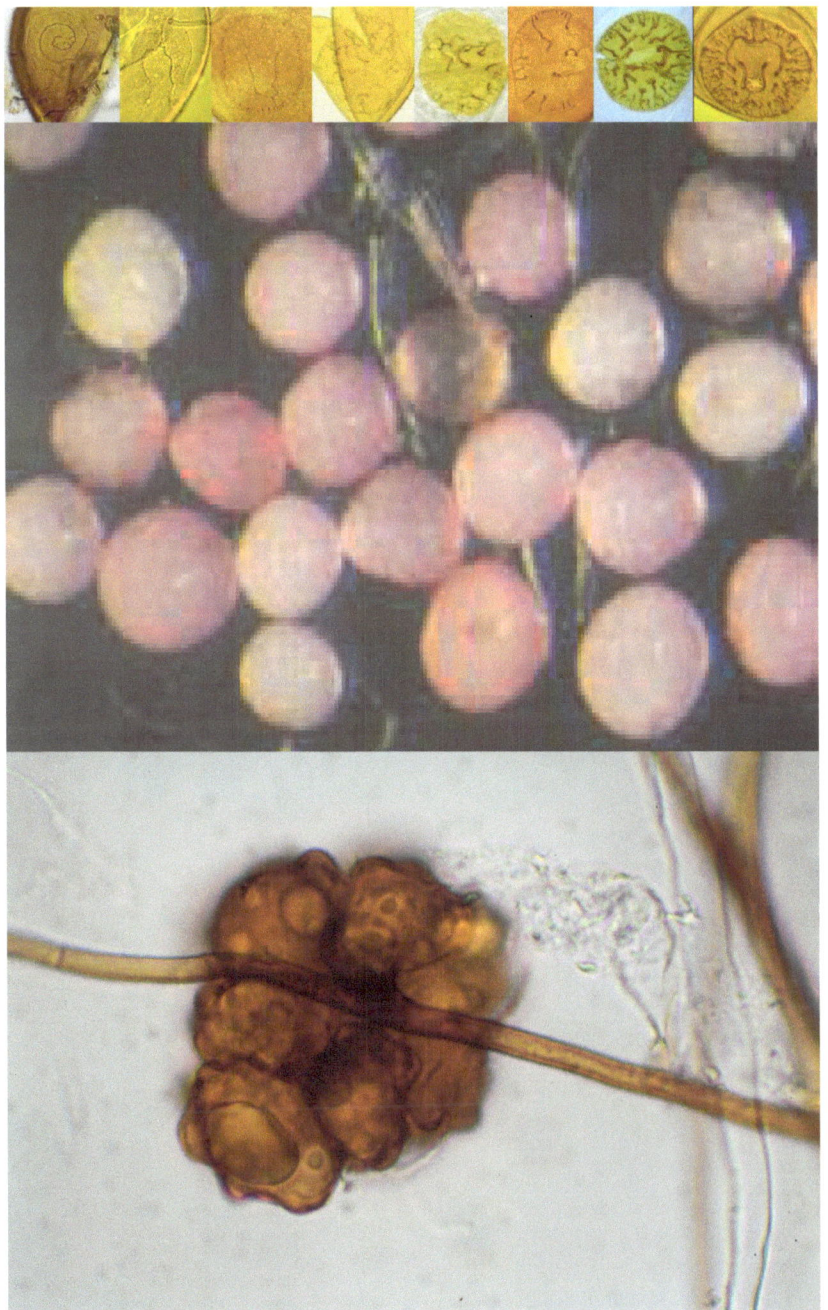

SCUTELLOSPORA

Genus name: *Scutellospora* C. Walker & F.E. Sanders, Mycotaxon 27: 179 (1986)

Included species

Generic type: *Scutellospora calospora* (T.H. Nicolson & Gerd.) C. Walker & F.E. Sanders, Mycotaxon 27: 180 (1986)

≡*Endogone calospora* T.H. Nicolson & Gerd., Mycologia 60(2): 322 (1968)

As far as is known, although this is the type species of the genus, all the early cultures, including those from which the type was sampled, are lost. The re-description of the species by Koske & Walker (1986) was made from a culture established from within 75 km of the type location, and its identity was confirmed by the late Prof. T. Nicolson. It is registered as BEG32, and we hereby formally designate it as an epitype.

Type: Scotland, Perthshire, East Newton, Dec 1966 **FH**.

Epitype: 5735, 20 Apr 2010, in the C. Walker collection in **E**, here designated. Derived from the culture with the designator Attempt 333-38 with Attempt 333-3, a single spore isolate of C. Walker, in its ancestry. Ex-epitype culture material is available from BEG (www.kent.ac.uk/bio/beg/englishhomepage.htm) as BEG32.

Scutellospora nodosa Błaszk., Mycologia 83(4): 537 (1991)

This species was defined from a field collection (Blaszkowski 1991) and as far as is known there is no culture available. Its phylogenetic position is anchored by a culture from the north of England that is registered as BEG4.

Type: Holotype: Hel, Poland, J. Błaszkowski 1491, 5 Aug 1989 **DPP** (Dept. Plant Pathology, Academy of Agriculture, Słowackiego, Szczecin, Poland): Isotypes 1472-1490 and 1492-1501, **DPP** and **OSC**.

Epitype: W3211, 27 Jan 1999, lodged in the C. Walker collection in **E**, here designated. From the culture with the designator Attempt 209-33 that had Attempt 209-5, a single spore isolate of C. Walker, in its ancestry. Ex-epitype culture material is available from BEG (www.kent.ac.uk/bio/beg/englishhomepage.htm) as BEG4.

Scutellospora projecturata Kramad. & C. Walker, in Kramadibrata, Walker, Schwarzott & Schüßler, Ann. Bot., Lond., N.S. 86(1): 22 (2000)

Scutellospora dipurpurescens J.B. Morton & Koske, Mycologia 80(4): 520 (1988)

Scutellospora aurigloba (I.R. Hall) C. Walker & F.E. Sanders, Mycotaxon 27: 180 (1986)

SPECIES OF UNCERTAIN POSITION IN *SCUTELLOSPORA*

The following do not have molecular evidence to place them in the genus *Scutellospora* sensu Goto & Oehl (2010) but are left in *Scutellospora* by Morton & Msiska (2010). We have retained or placed them in *Scutellospora* on the grounds that the morphological characters used by Oehl et al. (2008) were unsustainable (Msiska & Morton 2010) but we point out that molecular phylogenies clearly indicate that the species represent several genera.

Scutellospora arenicola Koske Koske & Halvorson, Mycologia 81(6): 927 (1990) [1989]

Scutellospora armeniaca Błaszk., Mycologia 84(6): 939 (1993) [1992]

≡*Cetraspora armeniaca* (Błaszk.) Oehl, F.A. Souza & Sieverd., Mycotaxon 106: 338 (2008)

Scutellospora biornata Spain, Sieverd. & S. Toro, Mycotaxon 35(2): 220 (1989)

≡*Dentiscutata biornata* (Spain, Sieverd. & S. Toro) Sieverd., F.A. Souza & Oehl, Mycotaxon 106: 342 (2008)

Scutellospora cerradensis Spain & J. Miranda, Mycotaxon 60: 130 (1996)

  ≡*Dentiscutata cerradensis* (Spain & J. Miranda) Sieverd., F.A. Souza & Oehl, Mycotaxon 106: 342 (2008)

  =*Scutellospora trirubiginopa* X.L. Pan & G.Yun Zhang, in Pan, Zhang, Wang & Wu, Mycosystema 16(3): 169 (1997)

  =*Fuscutata trirubiginopa* (X.L. Pan & G.Yun Zhang) Oehl, F.A. Souza & Sieverd., Mycotaxon 106: 347 (2008)

  Repeated attempts have been made to obtain the type of *Fuscutata trirubiginopa* for examination, but all have failed. The species description is such that it clearly is a synonym of *S. cerradensis*.

Scutellospora crenulata R.A. Herrera, Cuenca & C. Walker, Can. J. Bot. 79(6): 674 (2001)

Scutellospora dipapillosa (C. Walker & Koske) C. Walker & F.E. Sanders, Mycotaxon 27: 181 (1986)

  ≡*Gigaspora dipapillosa* C. Walker & Koske, in Koske & C. Walker 1985, Mycologia 77(5): 709 (1985)

Scutellospora erythropus (Koske & C. Walker) C. Walker & F.E. Sanders [as 'erythropa'], Mycotaxon 27: 181 (1986)

  ≡*Gigaspora erythropus* Koske & C. Walker [as 'erythropa'], Mycologia 76(2): 250 (1984)

  ≡*Quatunica erythropus* (Koske & C. Walker) F.A. Souza, Sieverd. & Oehl, in Oehl, Souza & Sieverding, Mycotaxon 106: 348 (2008)

Scutellospora gilmorei (Trappe & Gerd.) C. Walker & F.E. Sanders, Mycotaxon 27: 181 (1986)

  ≡*Gigaspora gilmorei* Trappe & Gerd., Mycol. Mem. 5: 27 (1974)

  ≡*Cetraspora gilmorei* (Trappe & Gerd.) Oehl, F.A. Souza & Sieverd., Mycotaxon 106: 338 (2008)

Scutellospora hawaiiensis Koske & Gemma, Mycologia 87(5): 678 (1995)

  ≡*Dentiscutata hawaiiensis* (Koske & Gemma) Sieverd., F.A. Souza & Oehl, Mycotaxon 106: 342 (2008)

Scutellospora heterogama (T.H. Nicolson & Gerd.) C. Walker & F.E. Sanders, Mycotaxon 27: 180 (1986)

  ≡*Endogone heterogama* T.H. Nicolson & Gerd., Mycologia 60(2): 319 (1968)

  ≡*Gigaspora heterogama* (T.H. Nicolson & Gerd.) Gerd. & Trappe, Mycol. Mem. 5: 31 (1974)

  ≡*Dentiscutata heterogama* (T.H. Nicolson & Gerd.) Sieverd., F.A. Souza & Oehl, Mycotaxon 106: 342 (2008)

Scutellospora nigra (J.F. Redhead) C. Walker & F.E. Sanders, Mycotaxon 27: 181 (1986)

  ≡*Gigaspora nigra* J.F. Redhead, Mycologia 71(1): 187 (1979)

  ≡*Dentiscutata nigra* (J.F. Redhead) Sieverd., F.A. Souza & Oehl, Mycotaxon 106: 342 (2008)

Scutellospora pellucida (T.H. Nicolson & N.C. Schenck) C. Walker & F.E. Sanders, Mycotaxon 27: 181 (1986)

  ≡*Gigaspora pellucida* T.H. Nicolson & N.C. Schenck, Mycologia 71(1): 189 (1979)

  ≡*Cetraspora pellucida* (T.H. Nicolson & N.C. Schenck) Oehl, F.A. Souza & Sieverd., Mycotaxon 106: 338 (2008)

Scutellospora pernambucana Oehl, Oehl, D.K. Silva, N. Freitas & L.C. Maia, Mycotaxon 106: 363 (2008)

Scutellospora reticulata (Koske, D.D. Mill. & C. Walker) C. Walker & F.E. Sanders, Mycotaxon 27: 181 (1986)

  ≡*Gigaspora reticulata* Koske, D.D. Mill. & C. Walker, Mycotaxon 16(2): 429 (1983)

  ≡*Dentiscutata reticulata* (Koske, D.D. Mill. & C. Walker) Sieverd., F.A. Souza & Oehl, Mycotaxon 106: 342 (2008)

Scutellospora rubra Stürmer & J.B. Morton, Mycol. Res. 103(8): 951 (1999)

  ≡*Fuscutata rubra* (Stürmer & J.B. Morton) Oehl, F.A. Souza & Sieverd., Mycotaxon 106: 347 (2008)

Scutellospora savannicola (R.A. Herrera & Ferrer) C. Walker & F.E. Sanders, Mycotaxon 27: 180 (1986)

≡*Gigaspora savannicola* R.A. Herrera & Ferrer, in Ferrer & Herrera, Revta Jardín Bot. Nac., Univ. Habana 1(1): 57 (1981) [1980]

≡*Fuscutata savannicola* (R.A. Herrera & Ferrer) Oehl, F.A. Souza & Sieverd., Mycotaxon 106: 347 (2008)

Scutellospora scutata C. Walker & Dieder., Mycotaxon 35(2): 357 (1989)

≡*Dentiscutata scutata* (C. Walker & Dieder.) Sieverd., F.A. Souza & Oehl, Mycotaxon 106: 342 (2008)

Scutellospora spinosissima C. Walker & Cuenca, in Walker, Cuenca & Sánchez, Ann. Bot., Lond., n.s. 82(6): 723 (1998)

≡*Cetraspora spinosissima* (C. Walker & Cuenca) Oehl, F.A. Souza & Sieverd., Mycotaxon 106: 340 (2008)

Scutellospora striata Cuenca & R.A. Herrera, Mycotaxon 105: 81 (2008)

≡*Cetraspora striata* (Cuenca & R.A. Herrera) Oehl, F.A. Souza & Sieverd., Mycotaxon 106: 340 (2008)

Scutellospora tricalypta (R.A. Herrera & Ferrer) C. Walker & F.E. Sanders, Mycotaxon 27: 180 (1986)

≡*Gigaspora tricalypta* R.A. Herrera & Ferrer, in Ferrer & Herrera, Revta Jardín Bot. Nac., Univ. Habana 1(1): 49 (1981) [1980]

RACOCETRA

This genus was separated along with another four putative genera (*Cetraspora, Dentiscutata, Quatunica* and *Fuscutata*) by Oehl et al (2008) from *Scutellospora*, and was the only generic name accepted in a review of that work by Msiska & Morton (2010). However, new molecular phylogentic data partly support the genera erected by Oehl et al. (2008). Our molecular analysis confirms that the genus *Cetraspora* must be revived to accommodate some species, if not synonymised with *Racocetra*, and it is likely that one or more of the other three will be needed to accommodate species in the *Scutellosporaceae*. Nevertheless, some of the species of *Scutellospora* were transferred to these genera on the basis of a faulty morphological reasoning without benefit of molecular evidence, and we have placed these in the list of *Scutellospora* species of uncertain position until such evidence is available.

Genus name: *Racocetra* Oehl, F.A. Souza & Sieverd., Mycotaxon: 334 (2009)

Included species

Generic type: *Racocetra coralloidea* (Trappe, Gerd. & I. Ho) Oehl, F.A. Souza & Sieverd., Mycotaxon 106: 336 (2008)

≡*Gigaspora coralloidea* Trappe, Gerd. & I. Ho, Mycol. Mem. 5: 30 (1974)

≡*Scutellospora coralloidea* (Trappe, Gerd. & I. Ho) C. Walker & F.E. Sanders, Mycotaxon 27: 181 (1986)

Racocetra castanea (C. Walker) Oehl, F.A. Souza & Sieverd., Mycotaxon 106: 336 (2008)

Racocetra fulgida (Koske & C. Walker) Oehl, F.A. Souza & Sieverd., Mycotaxon 106: 336 (2008)

Racocetra gregaria (N.C. Schenck & T.H. Nicolson) Oehl, F.A. Souza & Sieverd. (2008), Mycotaxon 106: 337 (2008)

Racocetra persica (Koske & C. Walker) Oehl, F.A. Souza & Sieverd., Mycotaxon 106: 337 (2008)

≡*Gigaspora persica* Koske & C. Walker, Mycologia 77(5): 708 (1985)

≡*Scutellospora persica* (Koske & C. Walker) C. Walker & F.E. Sanders, Mycotaxon 27: 181 (1986)

Racocetra verrucosa (Koske & C. Walker) Oehl, F.A. Souza & Sieverd., Mycotaxon 106: 337 (2008)

≡*Gigaspora verrucosa* Koske & C. Walker, Mycologia 77(5): 705 (1985)

≡*Scutellospora verrucosa* (Koske & C. Walker) C. Walker & F.E. Sanders, Mycotaxon 27: 181 (1986)

Racocetra weresubiae (Koske & C. Walker) Oehl, F.A. Souza & Sieverd., Mycotaxon 106: 337 (2008)

≡*Scutellospora weresubiae* Koske & C. Walker (1986), Mycotaxon 27: 224 (1986)

Although not established in pot culture, specimens collected from sand dunes in Argentina were clearly of this species, and their molecular analysis allowed it to be placed in its phylogenetic context. Msiska & Morton (2010) removed this species from *Racocetra*, but our analyses clearly indicate it to be in a monophyletic clade with *Racocetra*, therefore we place it back in that genus.

SPECIES OF UNCERTAIN POSITION IN *RACOCETRA*

The following do not have any molecular evidence to place them in the genus. They were placed in the genus on the grounds of morphological characters that have been criticised by Msiska & Morton (2010).

Racocetra alborosea (Ferrer & R.A. Herrera) Oehl, F.A. Souza & Sieverd., Mycotaxon 106: 336 (2008)

≡*Gigaspora alborosea* Ferrer & R.A. Herrera, Revta Jardín Bot. Nac., Univ. Habana 1(1): 55 (1981) [1980]

≡*Scutellospora alborosea* (Ferrer & R.A. Herrera) C. Walker & F.E. Sanders, Mycotaxon 27: 180 (1986)

Racocetra beninensis Oehl, Tchabi & Lawouin, Mycotaxon 110: 201 (2010)

Racocetra intraornata B.T. Goto & Oehl, Mycotaxon 109: 485 (2009)

Racocetra minuta (Ferrer & R.A. Herrera) Oehl, F.A. Souza & Sieverd., Mycotaxon 106: 337 (2008)

≡*Gigaspora minuta* Ferrer & R.A. Herrera, Revta Jardín Bot. Nac., Univ. Habana 1(1): 53 (1981) [1980]

≡*Scutellospora minuta* (Ferrer & R.A. Herrera) C. Walker & F.E. Sanders, Mycotaxon 27: 180 (1986)

SPECIES OF UNCERTAIN POSITION IN *GIGASPORACEAE*

Gigaspora lazzarii Montecchi, Ruini & G. Gross, Riv. Micol. 39(1): 27 (1996)

The type material and species description of this organism share none of the characteristics of any member of the Glomeromycota.

Fuscutata heterogama Oehl, F.A. Souza, L.C. Maia & Sieverd., Mycotaxon 106: 344 (2008)

According to the concept of Oehl et al (2008) this organism, used as the type species for the genus, is not conspecific with *Scutellospora heterogama* (*Dentiscutata heterogama* in their concept) and therefore most existing cultures formerly identified as *S. heterogama* should be represented by the proposed name *Fuscutata heterogama*. This was done without any molecular evidence, although the culture ex-type is cited as being available. It is possible that *Fuscutata heterogama* in future will be a synonym, but because it is described as the type species of a genus we list it here awaiting further molecular evidence.

Dentiscutata colliculosa B.T. Goto & Oehl Nova Hedwigia 90: 385 (2010)

The phylogenetic position of this species places it amongst the genera rejected by Msiska & Morton (2010), so we list it here pending clarification.

Dentiscutata nigerita S. W. Kade Mycosphere 1(3), 243 (2010)

The phylogenetic position of this species places it amongst the genera rejected by Msiska & Morton (2010), so we list it here pending clarification.

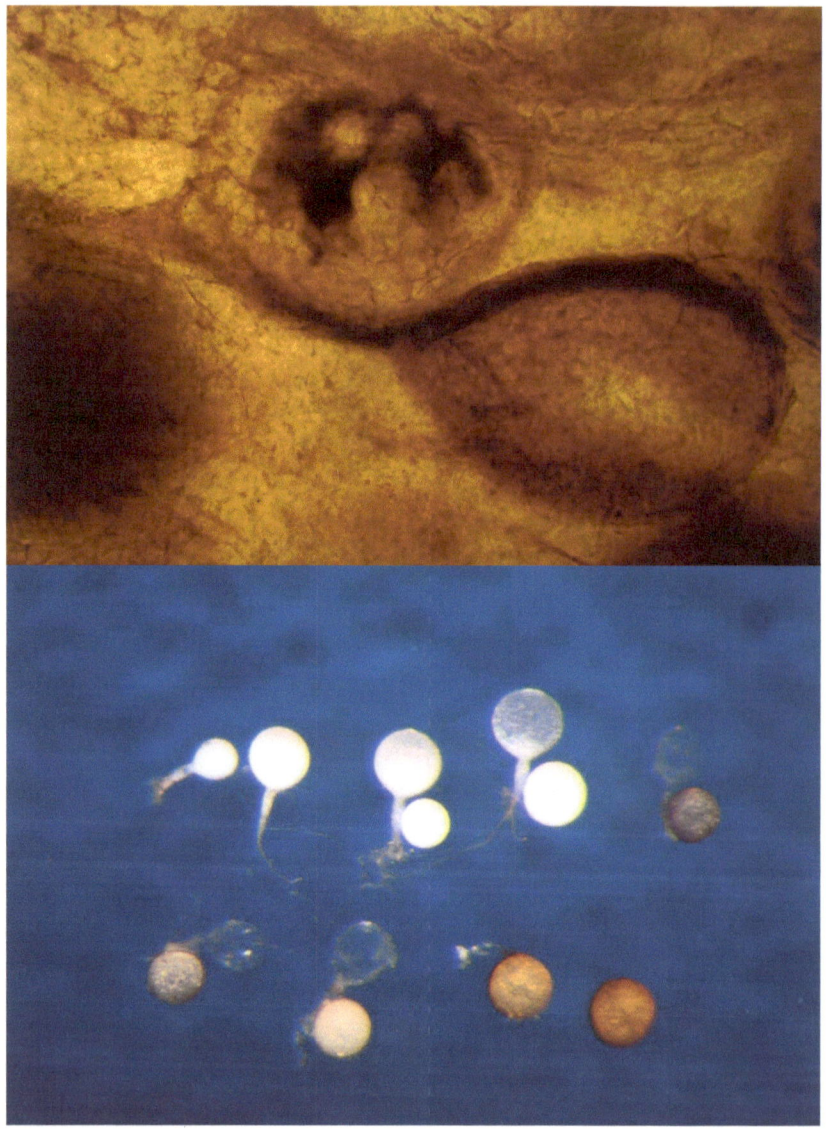

ACAULOSPORACEAE

Family name: *Acaulosporaceae* J.B. Morton & Benny, Mycotaxon 37: 479 (1990)

This family formerly included all the species in *Acaulospora* sensu lato, but of late, members of other groups (e.g., *Entrophospora*) have been moved into it (Kaonongbua 2010). Somewhat like *Glomus* sensu lato, for some time anything with an acaulosporoid spore was placed in *Acaulospora*, but the phylogenetic evidence has already been used to move such species as the former *Acaulospora trappei* into a natural classification (now *Archaeospora trappei*; Morton & Redecker 2001). However, there are at least two major clades contained within *Acaulospora*. At first it was thought these might divide naturally into ornamented and smooth-spored species, but this appears not to be the case. We do not yet have robust and sufficient molecular evidence to separate *Acaulospora* sensu lato into two genera, although it is most likely that it will be necessary to do so. As with the other genera, there are still many species for which no reliable molecular evidence is available, and these are listed as *Acaulospora* species of uncertain position.

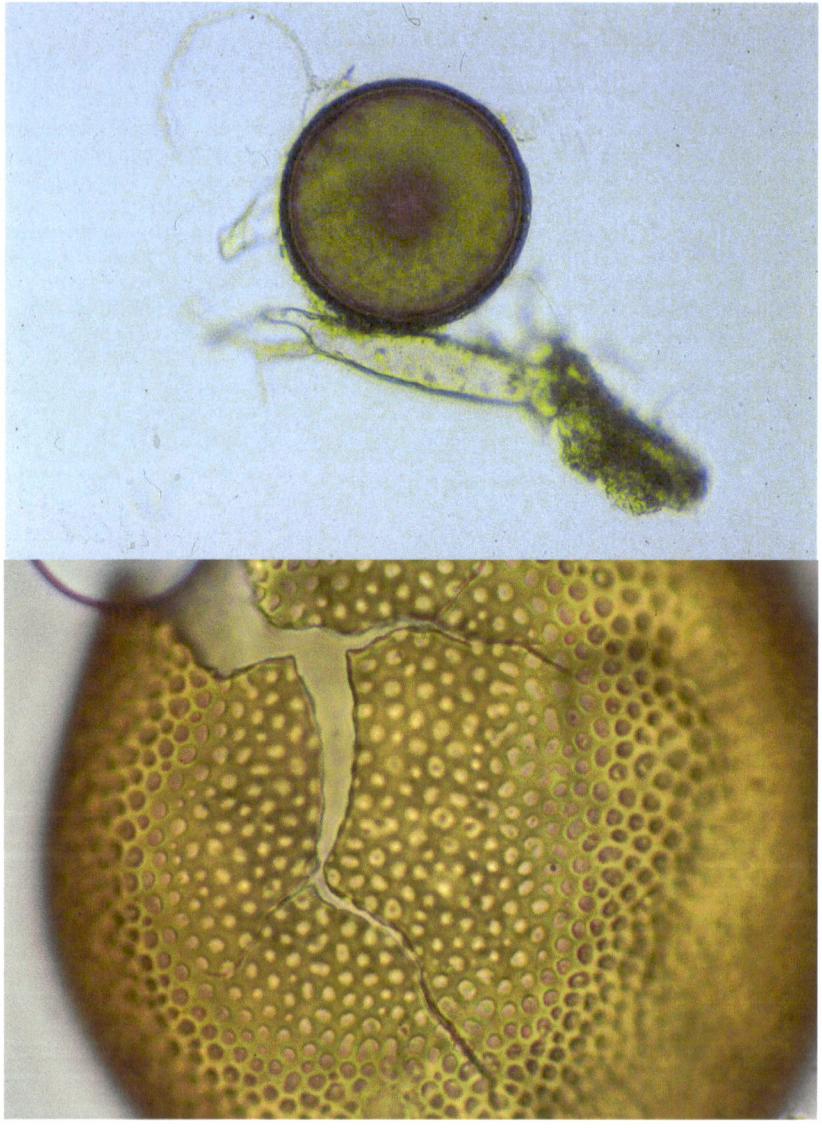

ACAULOSPORA

This is the type genus of the family. Its type species, *A. laevis*, was described from pot cultured specimens, and existed as ex-type cultures. Unfortunately all such cultures have been lost. We designate BEG13 as an epitype.

Genus name: *Acaulospora* Gerd. & Trappe, Mycol. Mem. 5: 31 (1974)

> =*Kuklospora* Oehl & Sieverd., J. Applied Bot. Food Quality 80: 74 (2006)

Included species

Generic type: *Acaulospora laevis* Gerd. & Trappe, Mycol. Mem. 5: 33 (1974)

> **Type**: Oregon, Benton Co., NE foot of Mary's Peak, Woods Creek, 6 Oct 1969, Trappe 2085 **OSC**.

> **Epitype**: W5258, 27 Jan 1999, in the C. Walker collection in **E**, here designated. Derived from the culture with the designator Attempt 192-10 that had Attempt 192-0, a multi spore culture made by B. Mosse at East Malling, known as the 'Rothamsted *A. laevis*', and also as the 'Honey coloured, sessile spore' in its ancestry); ex-epitype culture material is available from the BEG collection (www.kent.ac.uk/bio/beg/englishhomepage.htm) as BEG13.

Acaulospora mellea Spain & N.C. Schenck, in Schenck, Spain, Sieverding & Howeler, Mycologia 76(4): 689 (1984)

Acaulospora alpina Oehl, Sýkorová & Sieverd., in Oehl, Sýkorová, Redecker, Wiemken & Sieverding, Mycologia 98(2): 289 (2006)

> Classification based on molecular and morphological evidence from field material.

Acaulospora colliculosa Kaonongbua, J.B. Morton & Bever, Mycologia, 102(6): 1501-1503 (2010)

> This species is not available in culture, but the molecular analysis is made from authenticated specimens taken from the field. It is very close in spore morphology to *A. brasiliensis*, but separated from this species by sequences attributed to *A. alpina* (also from field collected spores).

Acaulospora colombiana (Spain & N.C. Schenck) Kaonongbua, J.B. Morton & Bever, Mycologia, 102(6): 1501 (2010)

> ≡*Entrophospora colombiana* Spain & N.C. Schenck, Mycologia 76: 693. 1984.

> ≡*Kuklospora colombiana* (Spain & N.C. Schenck) Oehl & Sieverd., J. Appl. Bot. Food Quality 80: 74. 2006.

Acaulospora entreriana M.S. Velázquez & Cabello, Mycotaxon 103: 179 (2008)

Acaulospora foveata Trappe & Janos, in Janos & Trappe, Mycotaxon 15: 516 (1982)

> The molecular evidence for this is taken from two INVAM cultures, CR315 and CR401A. The ornamentation illustrated for the species (for culture BR861) on the INVAM website is not entirely consistent with that of the type material for the species. The species description does not illustrate the full amount of variation seen in the spore ornamentation of the type material. It seems likely that spore ornamentation in these fungi is somewhat variable.

Acaulospora lacunosa J.B. Morton, Mycologia 78(4): 643 (1986)

Acaulospora cavernata Błaszk., Cryptog. Bot. 1(2): 204 (1989)

> **Type**: Holotype: Hel, Poland, J. Błaszkowski DPP 1285, 28 Sep 1988 **DPP**: Isotypes 1259-1267 and 1277-1267, 1277-1284, 1286-1298, 1386-1387 **DPP**.

> A culture, at first considered to be *Acaulospora scrobiculata*, was established from the United Kingdom, and registered as BEG33. The culture was established from five spores that all appeared, under the dissecting microscope, to have ornamentation similar to *A. scrobiculata* but the spores produced now are predominantly similar to the type of *A. cavernata*. The type of *A. cavernata* consists only of preserved spores on microscope slides, and thus is not available for molecular study. We therefore are making use of the BEG33 culture to designate an epitype for the species.

> **Epitype**: W5738, 30 Apr 2010, in the C. Walker collection in **E**, here designated. Derived from the culture with the designator Attempt 209-66 that had Attempt 209-1, a culture inoculated with five spores made by C. Walker in its ancestry). Ex-epitype culture material is available from BEG (www.kent.ac.uk/bio/beg/englishhomepage.htm) as BEG33.

Acaulospora delicata C. Walker, C.M. Pfeiff. & Bloss, Mycotaxon 25(2): 622 (1986)

Acaulospora dilatata J.B. Morton, Mycologia 78(4): 641 (1986)

Acaulospora kentinensis (C.G. Wu & Y.S. Liu) Kaonongbua, J.B. Morton & Bever, Mycologia, 102(6): 1501 (2010)

    ≡*Entrophospora kentinensis* C.G. Wu & Y.S. Liu, Mycotaxon 53: 287 (1995)

    ≡*Kuklospora kentinensis* (C.G. Wu & Y.S. Liu) Oehl & Sieverd., J. App. Bot. Food Quality 80:74 (2006)

Acaulospora koskei Błaszk., Mycol. Res. 99(2): 237 (1995)

Acaulospora longula Spain & N.C. Schenck, in Schenck, Spain, Sieverding & Howeler, Mycologia 76(4): 689 (1984)

Acaulospora morrowiae Spain & N.C. Schenck [as 'morrowae'], in Schenck, Spain, Sieverding & Howeler, Mycologia 76(4): 692 (1984)

Acaulospora paulinae Błaszk., Bulletin of the Polish Academy of Sciences, Biological Sciences 36(10-12): 273 (1988)

Acaulospora scrobiculata Trappe, Mycotaxon 6(2): 363 (1977)

Acaulospora spinosa C. Walker & Trappe, Mycotaxon 12(2): 515 (1981)

Acaulospora tuberculata Janos & Trappe, Mycotaxon 15: 519 (1982)

SPECIES OF UNCERTAIN POSITION IN *ACAULOSPORA*

Acaulospora bireticulata F.M. Rothwell & Trappe, Mycotaxon 8(2): 472 (1979)

> This species probably is a synonym for *A. elegans*. If spores of the latter are examined, their appearance can be changed from one species to the other by switching in or out of Nomarski differential interference contrast microscopy.

Acaulospora capsicula Błaszk., Mycologia 82(6): 794 (1990)

> This has many similarities with *A. laevis*.

Acaulospora colossica P.A. Schultz, Bever & J.B. Morton, Mycologia 91(4): 677 (1999)

> This is similar to *A. laevis*.

Acaulospora denticulata Sieverd. & S. Toro, Angew. Bot. 61(3-4): 217 (1987)

Acaulospora elegans Trappe & Gerd., Mycol. Mem. 5: 34 (1974)

> This species has a most interesting ornamentation that, at least at the level of light microscopy, is almost identical to that of *Scutellospora reticulata*. It has been found in Europe as well as in North America.

Acaulospora excavata Ingleby & C. Walker, Mycotaxon 50: 100 (1994)

> The ornamentation on this species is of large deep cavities, leading to the suspicion that it belongs with such organisms as *A. scrobiculata*.

Acaulospora gedanensis Błaszk., Karstenia 27(2): 38 (1988) [1987]

Acaulospora myriocarpa Spain, Sieverd. & N.C. Schenck, in Schenck, Spain & Sieverding, Mycotaxon 25(1): 112 (1986)

Acaulospora nicolsonii C. Walker, L.E. Reed & F.E. Sanders, Trans. Br. Mycol. Soc. 83(2): 360 (1984)

> This species probably should be in the genus *Ambispora*, but must remain in *Acaulospora* pending molecular and other evidence.

Acaulospora polonica Błaszk., Karstenia 27(2): 38 (1988) [1987]

Acaulospora rehmii Sieverd. & S. Toro, Angew. Bot. 61(3-4): 219 (1987)

Acaulospora rugosa J.B. Morton, Mycologia 78(4): 645 (1986)

Acaulospora splendida Sieverd., Chaverri & I. Rojas, Mycotaxon 33: 252 (1988)

> Although this species produces acaulosporoid spores, they do not seem to share characteristics with other members of the genus, but as far as known, there is no material suitable for molecular analysis, and the species therefore is best retained in its current genus. Moving it elsewhere without further evidence would be speculative.

Acaulospora sporocarpia S.M. Berch, Mycotaxon 23: 409 (1985)

Acaulospora taiwania H.T. Hu, Quarterly Journal of Chinese Forestry 21(2): 48 (1988)

> This is very unlikely to be an *Acaulospora* sp. In the species description, it is described as being produced in sporocarps. The illustrations of the species are of such poor quality that it is impossible to draw conclusions other than that it does not look much like an acaulospore producing fungus. The species description itself is not much more help, and the type material is lost. Consequently, this name should be considered as useless.

Acaulospora thomii Błaszk., Karstenia 27(2): 40 (1988) [1987]

Acaulospora undulata Sieverd., Angew. Bot. 62(5-6): 373 (1988)

> There is some mystery about the position of this species. It is said that it is fundamentally different from other species in the genus, and it has even been considered as a possible member of *Archaeospora*. One major consideration is the lack of reaction to Melzer's reagent, but this is not unknown among the ornamented, small-spored species in the group in which the reaction varies with the age and condition of the spores. Until the species is re-discovered and established in pot culture, recombining it within a different genus would be speculative.

Acaulospora walkeri Kramad. & Hedger, Mycotaxon 37: 73 (1990)

> Known only from field collected material, this species is very close in morphology to other species in the genus with small, smooth, yellow spores.

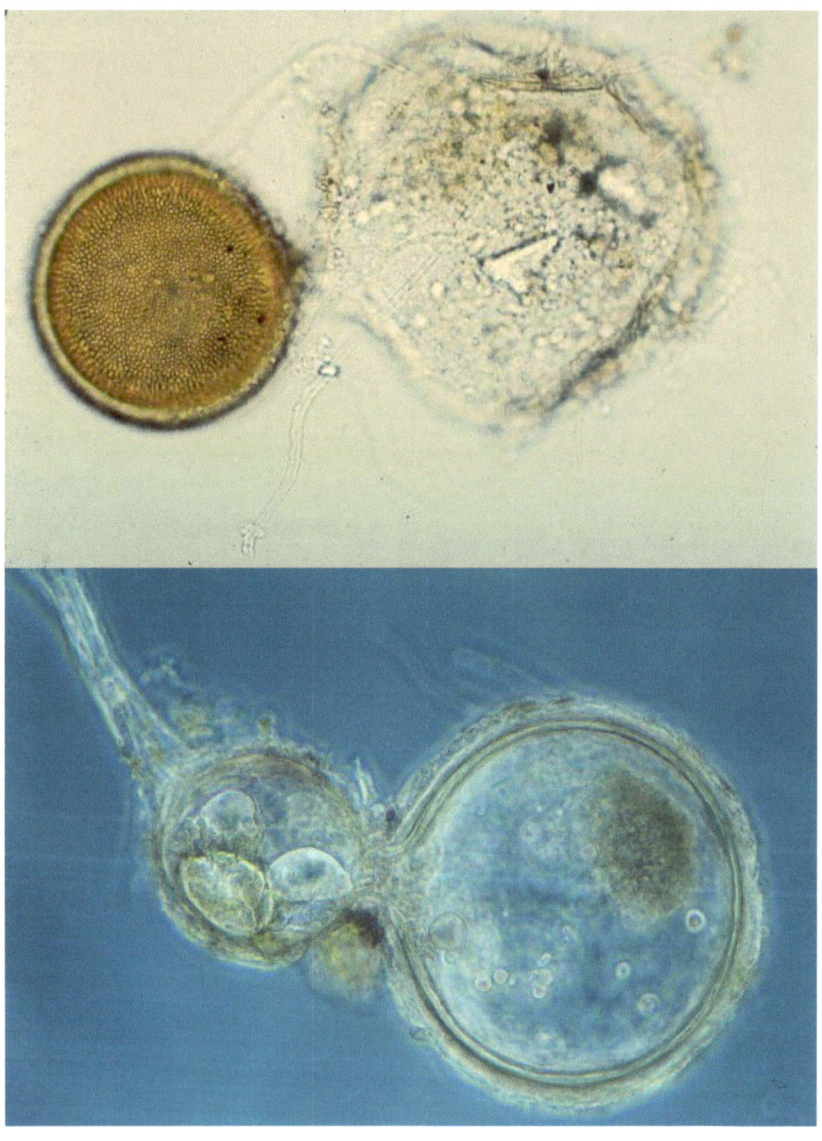

ENTROPHOSPORACEAE

This family (Sieverding & Oehl 2006) is of uncertain systematic position, because there seems to be no reliable evidence to substantiate it as natural. It will probably eventually be combined within others, but for now it is retained because there is no reliable evidence to place it separately or with any other group.

Family name: *Entrophosporaceae* Oehl & Sieverd., J. Appl. Bot. Food Quality (Angew. Botan.) 80: 73 (2006)

ENTROPHOSPORA

This genus is of uncertain systematic position. Because the family *Entrophosporaceae* was defined from *Entrophospora infrequens*, that species automatically belongs to the family unless and until the *Entrophospora* is formally revised, regardless of the fact that the molecular position is uncertain. One other species, *E. baltica*, is known only from field collections, and its correct phylogenetic position must remain a mystery pending further research.

Genus name: *Entrophospora* R.N. Ames & R.W. Schneid., Mycotaxon 8(2): 347 (1979)

Included species

Generic type: *Entrophospora infrequens* (I.R. Hall) R.N. Ames & R.W. Schneid., Mycotaxon 8(2): 348 (1979)

≡*Glomus infrequens* I.R. Hall, Trans. Br. Mycol. Soc. 68(3): 345 (1977)

SPECIES OF UNCERTAIN POSITION *ENTROPHOSPORA*

Entrophospora baltica Błaszk., Madej & Tadych, Mycotaxon 68: 167 (1998)

Entrophospora nevadensis J. Palenzuela, N. Ferrol & Oehl, Mycologia 102(3): 627 (2010)

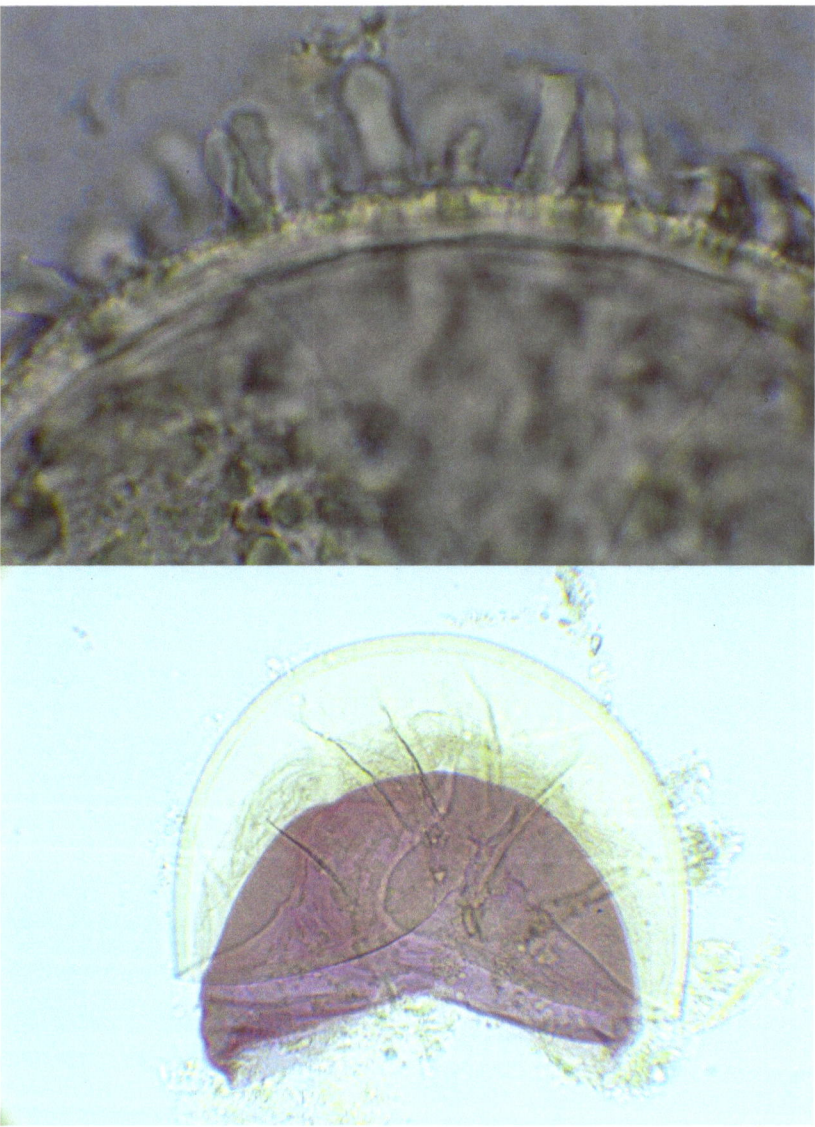

PACISPORACEAE

Family name: Pacisporaceae C. Walker, Błaszk., A. Schüßler & Schwarzott, in Walker & Schüßler, Mycol. Res. 108(9): 981 (2004)

This is another genus that was separated from *Glomus* sensu lato during the efforts to rationalise that genus. Its members produce spores with glomoid spores, but these spores have at least two wall groups, including at least one internal flexible wall component. It is said that some species form AM, but so far it seems they have proved impossible to maintain in pot culture for any length of time.

PACISPORA

Species are only based on field collected spores, and molecular evidence available for only *P. scintillans* and a fungus most likely corresponding to *P. franciscana* (unpublished data).

Genus name: Pacispora Oehl & Sieverd., J. Appl. Bot. (Angew. Bot.) 78: 74 (2004)

> =*Gerdemannia* C. Walker, Błaszk., A. Schüßler & Schwarzott, Mycol. Res. 108(6): 716 (2004)

Included species

Generic type: Pacispora chimonobambusae (C.G. Wu & Y.S. Liu) Sieverd. & Oehl ex C. Walker, Vestberg & A. Schüßler, in Walker, Vestberg & Schüßler, Mycol. Res. 111(3): 255 (2007)

> ≡*Gerdemannia chimonobambusae* (C.G. Wu & Y.S. Liu) C. Walker, Błaszk., A. Schüßler & Schwarzott, in Walker, Błaszkowski, Schwarzott & Schüßler, Mycol. Res. 108(6): 717 (2004)

> ≡*Glomus chimonobambusae* C.G. Wu & Y.S. Liu, in Wu, Liu, Hwuang, Wang & Chao, Mycotaxon 53: 284 (1995)

Pacispora boliviana Sieverd. & Oehl, in Oehl & Sieverding, J. Appl. Bot. (Angew. Bot.) 78: 79 (2004)

Pacispora coralloidae Sieverd. & Oehl, in Oehl & Sieverding, J. Appl. Bot. (Angew. Bot.) 78: 78 (2004)

Pacispora franciscana Sieverd. & Oehl, in Oehl & Sieverding, J. Appl. Bot. (Angew. Bot.) 78: 74 (2004)

Pacispora patagonica (Novas & Fracchia) C. Walker, Vestberg & A. Schüßler, Mycol. Res. 111(3): 255 (2007)

> ≡*Glomus patagonicum* Novas & Fracchia, Nova Hedwigia 80(3-4): 534 (2005)

Pacispora robigina Sieverd. & Oehl, in Oehl & Sieverding, J. Appl. Bot. (Angew. Bot.) 78: 75 (2004)

Pacispora scintillans (S.L. Rose & Trappe) Sieverd. & Oehl ex C. Walker, Vestberg & A. Schüßler, in Walker, Vestberg & Schüßler, Mycol. Res. 111(3): 255 (2007)

> ≡*Gerdemannia scintillans* (S.L. Rose & Trappe) C. Walker, Błaszk., A. Schüßler & Schwarzott, in Walker, Błaszkowski, Schwarzott & Schüßler, Mycol. Res. 108(6): 716 (2004)

> ≡*Glomus scintillans* S.L. Rose & Trappe, Mycotaxon 10(2): 417 (1980)

> =*Pacispora dominikii* (Błaszk.) Sieverd. & Oehl, in Oehl & Sieverding, J. Appl. Bot., Angew. Bot. 78: 76 (2004)

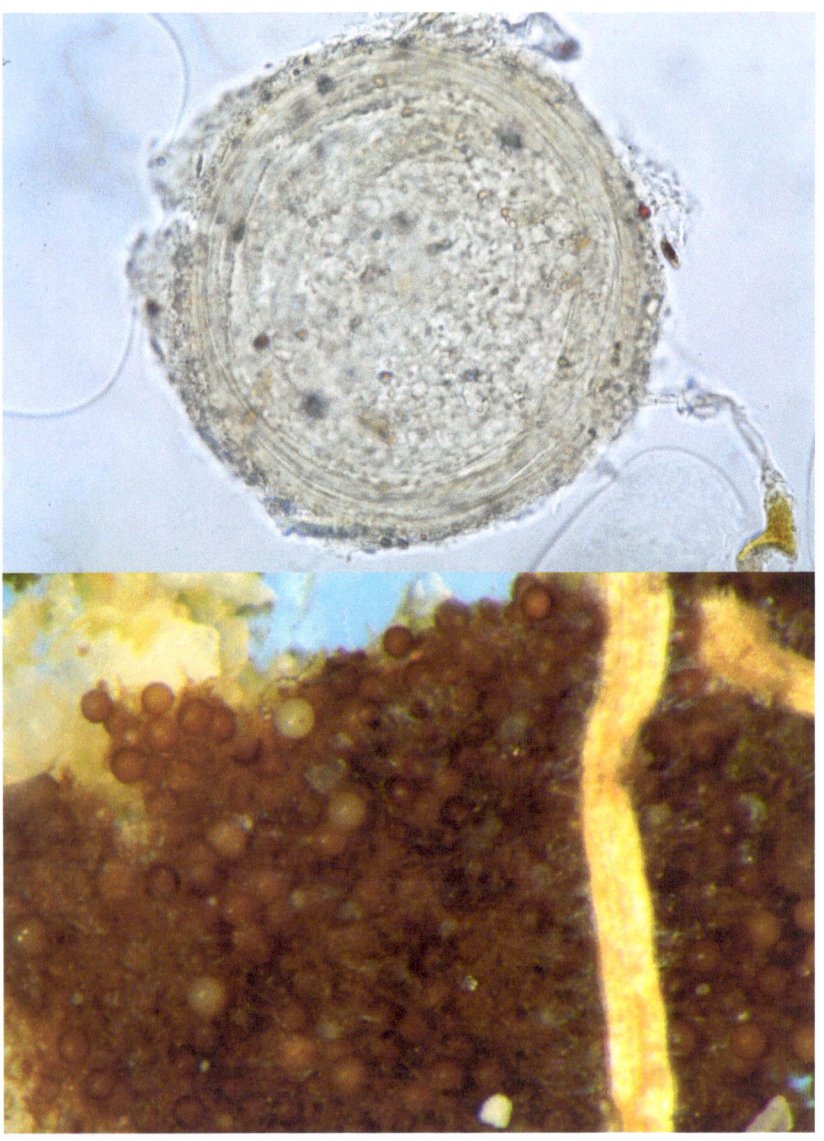

DIVERSISPORACEAE

This is a family based on the separation of a natural clade from *Glomus* sensu lato. *Glomus epigaeum*, published by Daniels & Trappe (1979) has been widely used as a 'model AMF'. It was synonymised with the former *Endogone versiformis* by Berch & Fortin (1983) but we have long since published evidence elsewhere showing its natural phylogenetic position (e.g., Schüßler et al. 2001a; Schwarzott et al. 2001; Gamper et al. 2009), and that it is not synonymous with *G. versiforme*. It is newly combined here in the genus *Diversispora*. Similarly the species *G. aurantium, G. eburneum, and G. trimurales* must be transferred to *Diversispora*.

Family name: ***Diversisporaceae*** C. Walker & A. Schüßler, in Walker & Schüßler, Mycol. Res. 108(9): 981 (2004)

DIVERSISPORA

A number of species in *Glomus* are transferred here based on molecular evidence.

Genus name: *Diversispora* C. Walker & A. Schüßler, Mycol. Res. 108(9): 982 (2004)

Included species

Generic type: ***Diversispora spurca*** (C.M. Pfeiff., C. Walker & Bloss) C. Walker & A. Schüßler [as *'spurcum'*], Mycol. Res. 108(9): 982 (2004)

Diversispora celata C. Walker, Gamper & A. Schüßler, in Gamper, Walker & Schüßler, New. Phytol.182: 497 (2009)

Diversispora epigaea (B.A. Daniels & Trappe) C. Walker & A. Schüßler comb. nov.

≡*Glomus epigaeum* B.A. Daniels & Trappe, Can. J. Bot. 57: 540 (1979)

Diversispora eburnea (L.J. Kenn., J.C. Stutz & J.B. Morton) C. Walker & A. Schüßler comb. nov.

≡*Glomus eburneum* L.J. Kenn., J.C. Stutz & J.B. Morton, Mycologia 91: 1084 (1999)

Diversispora aurantium (Błaszk., Blanke, Renker & Buscot) C. Walker & A. Schüßler comb. nov.

≡*Glomus aurantium* Błaszk., Blanke, Renker & Buscot, Mycotaxon 90: 540 (2004)

Diversispora trimurales (Koske & Halvorson) C. Walker & A. Schüßler comb. nov.

≡*Glomus trimurales* Koske & Halvorson, Mycologia 81: 930 (1989)

OTOSPORA

The genus has unclear phylogenetic affiliation though placed among *Diversispora* based on two non-overlapping short SSU sequences. Currently, this has to be accepted because it was published, but the type is un-interpretable due to parasitism and other degradation, and as yet there is no independent evidence is available to support its phylogenetic position.

Genus name: ***Otospora*** Palenz., Ferrol & Oehl, in Palenzuela, Ferrol, Boller, Azcón-Aguilar & Oehl, Mycologia 100: 297 (2008)

Included species

Generic type: ***Otospora bareae*** Palenz., Ferrol & Oehl [as *'bareai'*], in Palenzuela, Ferrol, Boller, Azcón-Aguilar & Oehl, Mycologia 100: 298 (2008)

REDECKERA

These species epithets were all published within the genus *Glomus* from field collected material (Redecker et al. 2007). However, molecular evidence from such field samples places them in a separate clade at the level of genus. Consequently, we raise a new genus to accommodate them.

Redeckera C. Walker & A. Schüßler gen. nov.

> A generibus ceteris in Glomeromycota combinatus sporophoro glomoideo et sequentio DNA differenti. Sporocarpia ireegulare formans. Mycorrhizas arbusculares formans. Sequentia typica acidi desoxyribonucleici monadis 'SSU' ribosomatum: ARKTYTGGKMGCGGYAACGTRA .

> Forming glomoid spores in relatively large sporocarps with a peridium. With the sequence ARKTYTGGKMGCGGYAACGTRA of the small subunit ribosomal RNA gene specific for the genus.

Typus: *Glomus megalocarpum* D. Redecker (2007)

Type material: Guadeloupe, Vieux-Fort, Ravine Blondeau, mesophile forest, on forest floor, leg. C. Lécuru, 5.9. 2005. Holotype CL/Guad05-051 (LIP) (see Redecker et al. 2007).

Etymology: named in recognition of the pioneering work of Dirk Redecker in molecular phylogeny of the *Glomeromycota*

Included species

Generic type: *Redeckera megalocarpum* (D. Redecker) C. Walker & A. Schüßler comb. nov.

> ≡*Glomus megalocarpum* D. Redecker, in Redecker, Raab, Oehl, Camacho & Courtecuisse, Mycol. Prog. 6: 38 (2007)

Redeckera pulvinatum (Henn.) C. Walker & A. Schüßler comb. nov.

> ≡*Glomus pulvinatum* (Henn.) Trappe & Gerd. [as '*pulvinatus*'], in Gerdemann & Trappe, Mycol. Mem. 5: 59 (1974)

> ≡*Endogone pulvinata* Henn., Hedwigia 36: 212 (1897)

Redeckera fulvum (Berk. & Broome) C. Walker & A. Schüßler comb. nov.

> ≡*Glomus fulvum* (Berk. & Broome) Trappe & Gerd. [as '*fulvus*'], in Gerdemann & Trappe, Mycol. Mem. 5: 59 (1974)

> ≡*Paurocotylis fulva* Berk. & Broome, J. Linn. Soc., Bot. 14(2): 137 (1875)

> ≡*Endogone fulva* (Berk. & Broome) Pat., Bull. Soc. Mycol. Fr. 19: 341 (1903)

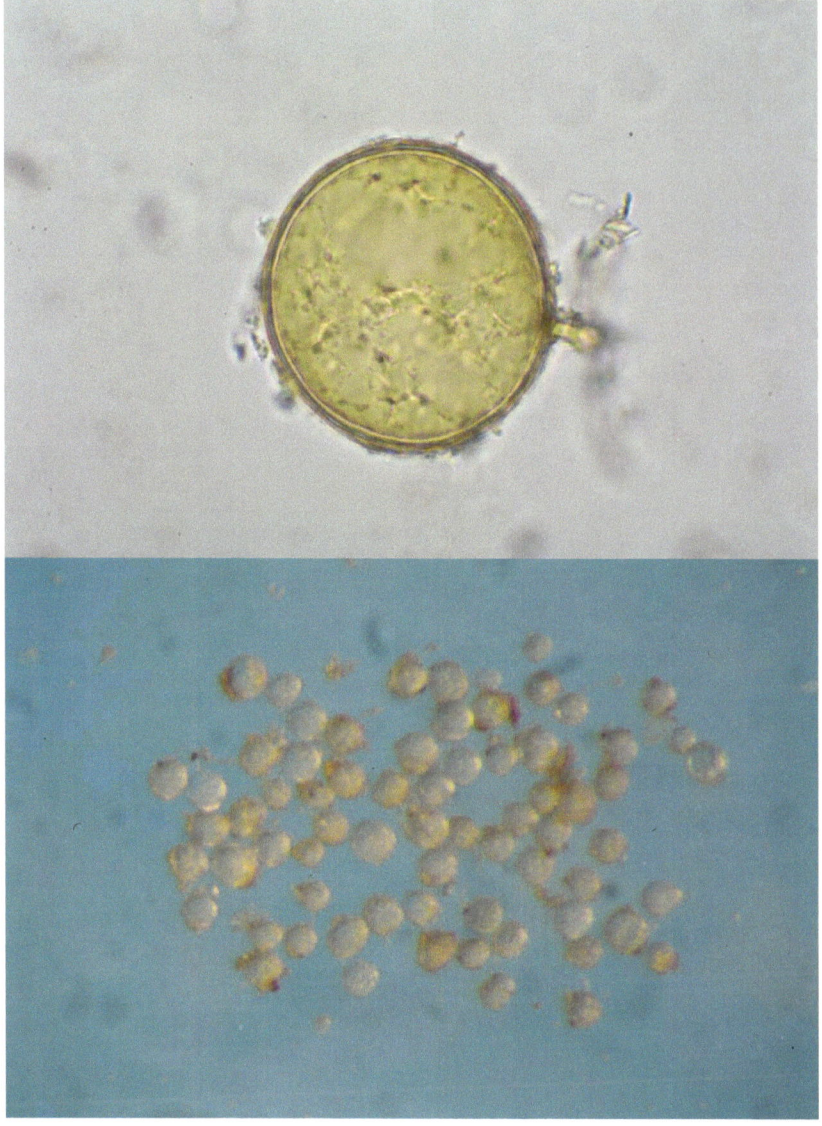

PARAGLOMERALES

Order entirely based on molecular evidence

Order name: *Paraglomerales* C. Walker & A. Schüßler, in Schüßler, Schwarzott & Walker, Mycol. Res. 105(12): 1418 (2001)

PARAGLOMERACEAE

Family entirely based on molecular evidence

Family name: *Paraglomeraceae* J.B. Morton & D. Redecker [as *'Paraglomaceae'*], Mycologia 93(1): 188 (2001)

PARAGLOMUS

This genus is based on molecular evidence that shows the spores, although appearing to be *Glomus*, are phylogenetically very distant from other members in that group.

Genus name: *Paraglomus* J.B. Morton & D. Redecker, Mycologia 93(1): 188 (2001)

Included species

Generic type: *Paraglomus occultum* (C. Walker) J.B. Morton & D. Redecker, Mycologia 93(1): 190 (2001)

≡ *Glomus occultum* C. Walker, Mycotaxon 15: 50 (1982)

Paraglomus brasilianum (Spain & J. Miranda) J.B. Morton & D. Redecker, Mycologia 93(1): 190 (2001)

≡ *Glomus brasilianum* Spain & J. Miranda, Mycotaxon 60: 139 (1996)

Paraglomus laccatum (Błaszk.) Renker, Błaszk. & Buscot, Nova Hedwigia 84(3-4): 400 (2007)

≡ *Glomus laccatum* Błaszk., Bulletin of the Polish Academy of Sciences, Biological Sciences 36(10-12): 271 (1988)

ARCHAEOSPORALES

Order name: *Archaeosporales* C. Walker & A. Schüßler, in Schüßler, Schwarzott & Walker, Mycol. Res. 105(12): 1418 (2001)

This is another group that was confounded because it produced species either with glomoid or acaulosporoid spores, or both. When molecular evidence became available, it became possible to separate them from their incorrect placements in *Glomus*. Included here is a most interesting organism, *Geosiphon pyriformis*, which so far forms the only known symbiosis between a member of the *Glomeromycota* and cyanobacteria of the genus *Nostoc*. Only one species is know, *Geosiphon pyriformis*.

GEOSIPHONACEAE

Family name: *Geosiphonaceae* Engl. & E. Gilg, Syllabus, Edn 9 & 10 (Berlin): 24 (1924)

GEOSIPHON

Genus name: *Geosiphon* F. Wettst., Öst. Bot. Z. 65: 152 (1915)

=*Geosiphonomyces* Cif. & Tomas., Atti Ist. Bot. Univ. Lab. Crittog. Pavia, Ser. 5 14: 5 (1957)

Included species

Generic type: *Geosiphon pyriformis* (Kütz.) F. Wettst. [as '*pyriforme*'], Öst. Bot. Z. 65: 147 (1915) emend Schüßler

≡*Botrydium pyriforme* Kütz., Spec. Alg.: 486 (1849)

≡*Geosiphonomyces pyriformis* Cif. & Tomas., Atti Ist. Bot. Univ. Lab. Crittog. Pavia, Ser. 5 14: 5 (1957)

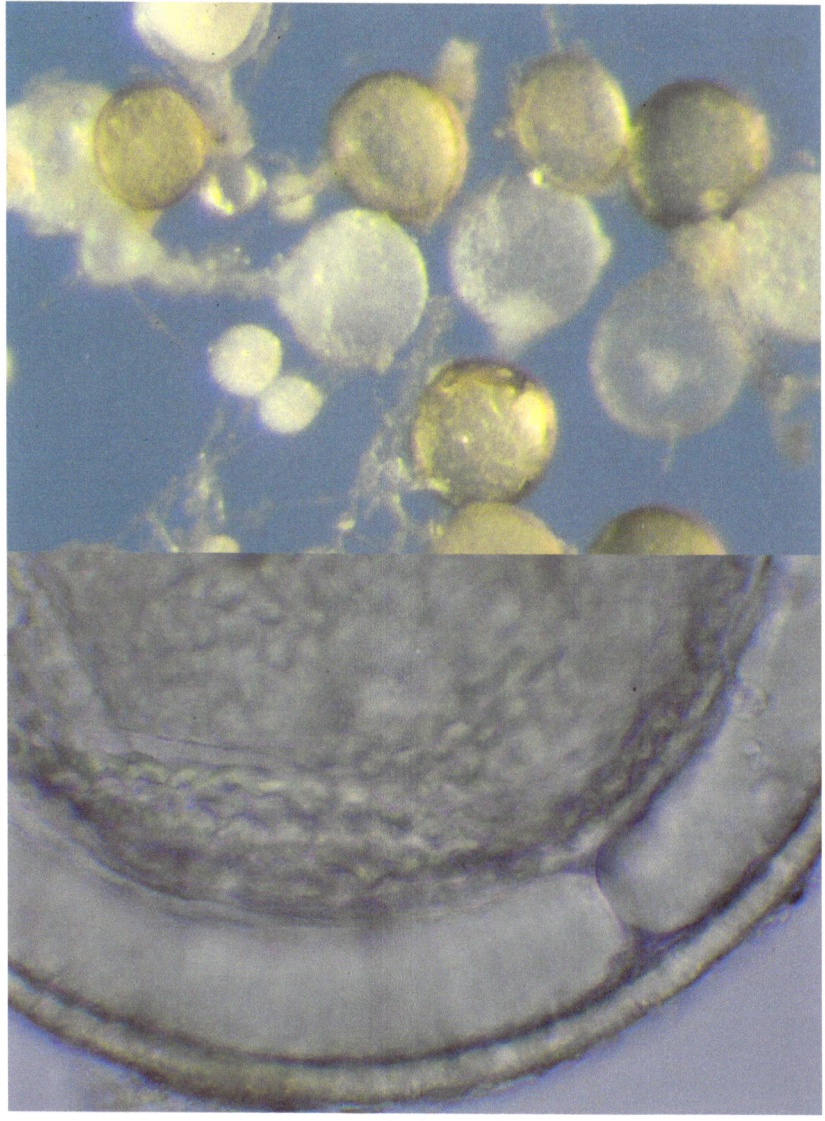

AMBISPORACEAE

Family name: *Ambisporaceae* C. Walker, Vestberg & A. Schüßler, in Walker, Vestberg, Demircik, Stockinger, Saito, Sawaki, Nishmura & Schüßler, Mycol. Res. 111(2): 143 (2007)

AMBISPORA

After it was found that an AM fungal species formed both acaulosporoid and glomoid spores, the molecular evidence was examined. It showed such organisms belonged to neither *Acaulospora* nor *Glomus* (Morton & Redecker 2001). A new genus, *Ambispora*, was later published (Walker et al. 2007), when it became evident that some species formerly in *Glomus*, *Acaulospora* and *Archaeospora* actually belonged here.

Genus name: *Ambispora* C. Walker, Vestberg & A. Schüßler, in Walker, Vestberg, Demircik, Stockinger, Saito, Sawaki, Nishmura & Schüßler, Mycol. Res. 111(2): 147 (2007)

Included species

Generic type: *Ambispora fennica* C. Walker, Vestberg & A. Schüßler, in Walker, Vestberg, Demircik, Stockinger, Saito, Sawaki, Nishmura & Schüßler, Mycol. Res. 111(2): 148 (2007)

Ambispora appendicula (Spain, Sieverd. & N.C. Schenck) C. Walker, Mycol. Res. 112(3): 298 (2008)

≡*Acaulospora appendicula* Spain, Sieverd. & N.C. Schenck, in Schenck, Spain, Sieverding & Howeler, Mycologia 76(4): 686 (1984)

≡*Appendicispora appendicula* (Spain, Sieverd. & N.C. Schenck) Spain, Oehl & Sieverd., in Spain, Sieverding & Oehl, Mycotaxon 97: 170 (2006)

≡*Paracaulospora appendicula* (Spain, Sieverd. & N.C. Schenck) S.P. Gautam & U.S. Patel, The Mycorrhizae, Diversity, Ecology and Applications (Delhi): 5 (2007)

Ambispora brasiliensis B.T. Goto, L.C. Maia & Oehl, Mycotaxon 105: 13 (2008)

Ambispora callosa (Sieverd.) C. Walker, Vestberg & A. Schüßler, in Walker, Vestberg, Demircik, Stockinger, Saito, Sawaki, Nishmura & Schüßler, Mycol. Res. 111(2): 148 (2007)

≡*Glomus callosum* Sieverd., Angew. Bot. 62(5-6): 374 (1988)

≡*Appendicispora callosa* (Sieverd.) C. Walker, Vestberg & A. Schüßler, Mycol. Res. 111(3): 254 (2007)

Ambispora fecundispora (N.C. Schenck & G.S. Sm.) C. Walker, Mycol. Res. 112(3): 298 (2008)

≡*Glomus fecundisporum* N.C. Schenck & G.S. Sm., Mycologia 74(1): 81 (1982)

≡*Appendicispora fecundispora* (N.C. Schenck & G.S. Sm.) C. Walker, Vestberg & A. Schüßler, Mycol. Res. 111(3): 254 (2007)

Ambispora gerdemannii (S.L. Rose, B.A. Daniels & Trappe) C. Walker, Vestberg & A. Schüßler, in Walker, Vestberg, Demircik, Stockinger, Saito, Sawaki, Nishmura & Schüßler, Mycol. Res. 111(2): 148 (2007)

≡*Glomus gerdemannii* S.L. Rose, B.A. Daniels & Trappe, Mycotaxon 8(1): 297 (1979)

≡*Archaeospora gerdemannii* (S.L. Rose, B.A. Daniels & Trappe) J.B. Morton & D. Redecker, Mycologia 93(1): 186 (2001)

≡*Appendicispora gerdemannii* (S.L. Rose, B.A. Daniels & Trappe) Spain, Oehl & Sieverd., in Spain, Sieverding & Oehl, Mycotaxon 97: 174 (2006)

Ambispora granatensis J. Palenzuela, N. Ferrol & Oehl in Palenzuela, Barea, Ferrol & Oehl, Mycologia 103, published online on 17 Oct 2010 as doi:10.3852/09-146

Ambispora jimgerdemannii (Spain, Oehl & Sieverd.) C. Walker, Mycol. Res. 112(3): 298 (2008)

≡*Appendicispora jimgerdemannii* Spain, Oehl & Sieverd., in Spain, Sieverding & Oehl, Mycotaxon 97: 174 (2006)

≡*Acaulospora gerdemannii* N.C. Schenck & T.H. Nicolson, Mycologia 71(1): 193 (1979)

Ambispora leptoticha (N.C. Schenck & G.S. Sm.) C. Walker, Vestberg & A. Schüßler, in C. Walker, Vestberg, Demircik, Stockinger, Saito, Sawaki, Nishmura & Schüßler, Mycol. Res. 111(2): 148 (2007)

≡*Glomus leptotichum* N.C. Schenck & G.S. Sm., Mycologia 74(1): 82 (1982)

≡*Appendicispora leptoticha* (N.C. Schenck & G.S. Sm.) C. Walker, Vestberg & A. Schüßler, Mycol. Res. 111(3): 255 (2007)

≡*Archaeospora leptoticha* (N.C. Schenck & G.S. Sm.) J.B. Morton & D. Redecker, Mycologia 93(1): 184 (2001)

≡*Pseudoglomus leptotichum* (N.C. Schenck & G.S. Sm.) S.P. Gautam & U.S. Patel, The Mycorrhizae, Diversity, Ecology and Applications (Delhi): 10 (2007)

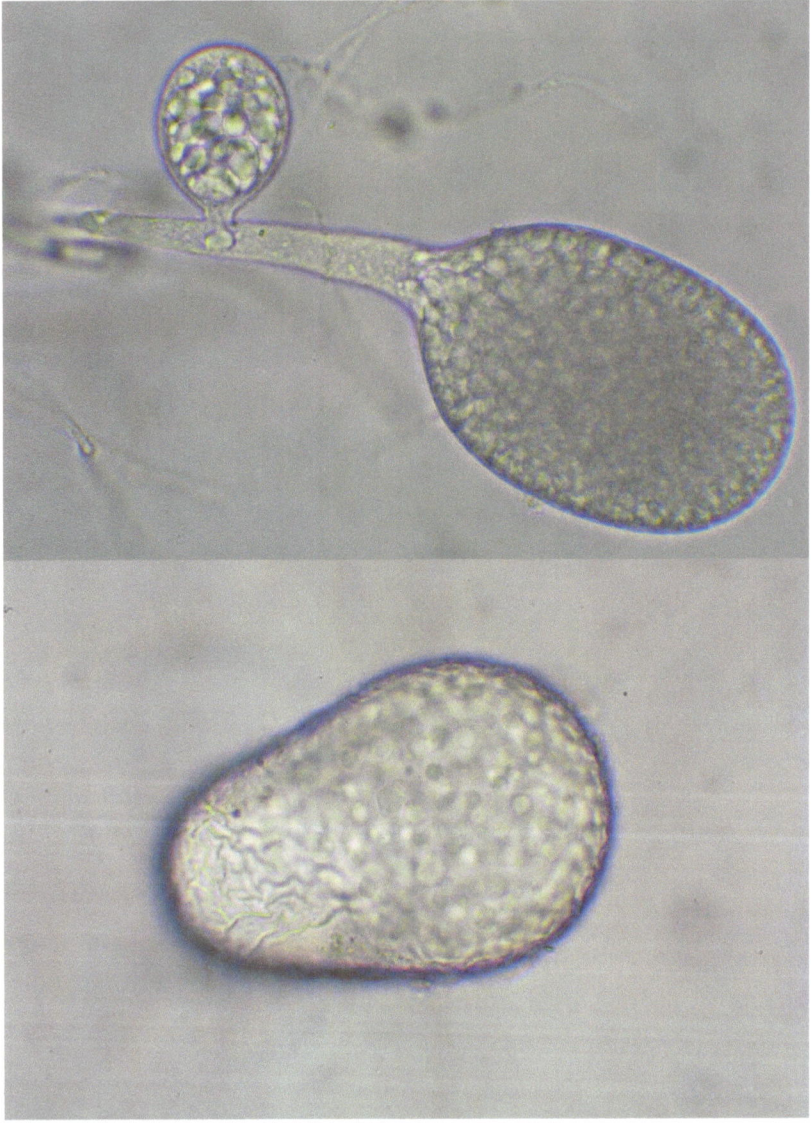

ARCHAEOSPORACEAE

Family name: *Archaeosporaceae* J.B. Morton & D. Redecker, Mycologia 93(1): 182 (2001)

ARCHAEOSPORA

Based on SSU sequences evidence *Intraspora schenkii* phylogenetically is placed in between different *Archaeospora trappei* cultures (which, however, might be non-conspecific) and shows low genetic distance to *Archaeospora*. Its morphologally also indicates that it is congeneric with *Archaeospora*. The situation is similar to the genus *Kuklospora* that was separated solely based on the orientation of formation of spores regarding the sporogenous saccule (Sieverding & Oehl 2006). This is not a reliable character for separating organisms in the *Glomeromycota*, as has been shown in the *Acaulosporaceae* and in the *Archaeosporaceae* (Morton & Redecker 2001; Kaonongbua 2010). We therefore recombine the species in *Archaeospora*.

Genus name: *Archaeospora* J.B. Morton & D. Redecker, Mycologia 93(1): 183 (2001)

>   =*Intraspora* Oehl & Sieverd., in Sieverding & Oehl, J. Appl. Bot. Food Quality (Angew. Botan.) 80: 77 (2006)

Included species

Generic type: *Archaeospora trappei* (R.N. Ames & Linderman) J.B. Morton & D. Redecker, Mycologia 93(1): 183 (2001)

Archaeospora schenckii (Sieverd. & S. Toro) C. Walker & A. Schüßler comb. nov.

>   ≡*Entrophospora schenckii* Sieverd. & S. Toro, Mycotaxon 28(1): 210 (1987)

>   ≡*Intraspora schenckii* (Sieverd. & S. Toro) Oehl & Sieverd., J. Appl. Bot. Food Quality (Angew. Botan.) 80: 77 (2006)

LITERATURE CITED

Almeida, RT, Schenck NC (1990) A revision of the genus *Sclerocystis* (*Glomaceae*: *Glomales*). Mycologia 82: 703-714

Berch SM, Fortin JA (1983) Lectotypification of *Glomus macrocarpum* and proposal of new combinations: *Glomus australe*, *Glomus versiforme*, and *Glomus tenebrosum* (*Endogonaceae*). Canadian Journal of Botany 61: 2608-2617

Berkeley MJ, Broome CE (1873) Enumeration of the fungi of Ceylon. Part II. Journal of the Linnean Society of London Botany 14: 137

Błaszkowski J (1991) Polish *Glomales* VIII. *Scutellospora nodosa*, a new species with knobby spores. Mycologia 83: 537-542

Chellappan P, Mohan N, Ramesh C, Selvaraj T, Arunachalam M, Mehadevan A (2005) A new approach for enhanced multiplication of arbuscular mycorrhizal fungi and isolation of ITS regions from *Glomus deserticola* and *Laccaria fraternal*. Indian Journal of Experimental Biology 43: 808-812

Daniels BA, Trappe JM (1979) *Glomus epigaeus* sp. nov., a useful fungus for vesicular-arbuscular mycorrhizal research. Canadian Journal of Botany 57: 539-542

Fonseca HMAC, Barbara RLL (2008) Does *Lunularia cruciata* form symbiotic relationships with either *Glomus proliferum* or *G. intraradices*? Mycological Research 112: 1063–1068

Gamper, HA, Walker C, Schuessler A (2009) *Diversispora celata* sp. nov.: molecular ecology and phylotaxonomy of an inconspicuous arbuscular mycorrhizal fungus. New Phytologist 182: 495-506

Gerdemann JW, Trappe JM (1974) The *Endogonaceae* in the Pacific Northwest. Mycologia Memoir No. 5: 1-76

Kaonongbua W, Morton JB, Bever JD (2010) Taxonomic revision transferring species in *Kuklospora* to *Acaulospora* (*Glomeromycota*) and a description of *Acaulospora colliculosa* sp. nov. from field collected spores. Mycologia, 102: 1497–1509

Koske RE, Walker C (1986) Species of *Scutellospora* (*Endogonaceae*) with smooth-walled spores from maritime sand dunes: two new species and a redescription of the spores of *Scutellospora pellucida* and *Scutellospora calospora*. Mycotaxon 27: 219-235

Krüger M, Stockinger H, Krüger C, Schüßler A (2009) DNA-based species level detection of *Glomeromycota*: one PCR primer set for all arbuscular mycorrhizal fungi. New Phytologist 183: 212-223

Morton JB, Franke M, Bentivenga SP (1998) Developmental foundations for morphological diversity among endomycorrhizal fungi in *Glomales* (*Zygomycetes*) In: Varma A, Hock B (eds). Mycorrhiza. Structure, Function, Molecular Biology and Biotechnology (2nd edn.). Springer-Verlag, Berlin Heidelberg New York

Morton JB, Msiska Z (2010) Phylogenies from genetic and morphological characters do not support a revision of *Gigasporaceae* (*Glomeromycota*) into four families and five genera. Mycorrhiza, 20: 483-496

Morton JB, Redecker D (2001) Two new families of *Glomales, Archaeosporaceae* and *Paraglomeraceae*, with two new genera *Archaeospora* and *Paraglomus*, based on concordant molecular and morphological characteristics. Mycologia 93: 181-195

Msiska Z, Morton JB (2008) Phylogenetic analysis of the *Glomeromycota* by partial β-tubulin gene sequences. Mycorrhiza 19(4): 247 - 254

Oehl F, de Souza FA, Sieverding E (2008) Revision of *Scutellospora* and descripton of five new genera and three new families in the arbuscular-forming *Glomeromycetes*. Mycotaxon 106: 311-360

Redecker D, Morton JB, Bruns TD (2000) Molecular phylogeny of the arbuscular mycorrhizal fungi *Glomus sinuosum* and *Sclerocystis coremiodes*. Mycologia 92: 282-285

Redecker D, Raab P, Oehl F, Camacho FJ, Courtecuisse R (2007) A novel clade of sporocarps-forming species of glomeromycotan fungi in the *Diversisporales* lineage. Mycological Progress 6: 35-44

Schüßler A (2000) *Glomus claroideum* forms an arbuscular mycorrhiza-like symbiosis with the hornwort *Anthoceros punctatus*. Mycorrhiza 10: 15-21

Schüßler A (2002) Molecular phylogeny, taxonomy, and evolution of *Geosiphon pyriformis* and arbuscular mycorrhizal fungi. Plant and Soil 244: 75-83

Schüßler A, Gehrig H, Schwarzott D, Walker C (2001) Analysis of partial *Glomales* SSU rRNA genes: implications for primer design and phylogeny. Mycological Research 105: 5-15

Schüßler A, Krüger M, Walker C (2011) Evolution of the 'plant-symbiotic' fungal phylum, *Glomeromycota*. In: The Mycota XIV - Evolution of Fungi and Fungal-like Organisms. Springer-Verlag, in press

Schüßler A, Schwarzott D, Walker C (2001b) A new fungal phylum, the *Glomeromycota*: phylogeny and evolution. Mycological Research 105: 1413-1421

Schwarzott D, Walker C, Schüßler A (2001) *Glomus*, the largest genus of the arbuscular mycorrhizal fungi (*Glomales*), is non-monophyletic. Molecular Phylogenetics and Evolution 21: 190-197

Sieverding W, Oehl F (2006) Revision of *Entrophospora* and description of *Kuklospora* and *Intraspora*, two new genera in the arbuscular mycorrhizal *Glomeromycetes*. Journal of Applied Botany and Food Quality 80: 69-81

Stockinger H, Walker C, Schüßler A 2009. '*Glomus intraradices* DAOM197198', a model fungus in arbuscular mycorrhiza research, is not *Glomus intraradices*. New Phytologist 183: 1176-1187

Tulasne LR, Tulasne C (1845) Fungi nonnulli hypogaei, novi. v. minus cogniti act. Giornale Botatico Italiano 2: 35-63

Walker C (2008) *Ambispora* and *Ambisporaceae* resurrected. Mycological Research 112: 297-298

Walker C, Błaszkowski J, Schwarzott D, Schüßler A (2004) *Gerdemannia* gen. nov., a genus separated from *Glomus*, and *Gerdemanniaceae* fam. nov., a new family in the *Glomeromycota*. Mycological Research 108: 707-718

Walker C, Vestberg M (1998) Synonymy amongst the arbuscular mycorrhizal fungi: *Glomus claroideum, G. maculosum, G. multisubstenum* [*sic*] and *G. fistulosum*. Annals of Botany82: 601-624

Walker C, Vestberg M, Demircik F, Stockinger H, Saito M, Sawaki H, Nishmura I, Schüßler A (2007) Molecular phylogeny and new taxa in the *Archaeosporales* (*Glomeromycota*): *Ambispora fennica* gen. sp. nov., *Ambisporaceae* fam. nov., and emendation of *Archaeosporaceae*. Mycological Research 111: 137-153

www.ingramcontent.com/pod-product-compliance
Lightning Source LLC
Chambersburg PA
CBHW040853180526
45159CB00001B/410